T0200152

Of Popes and Unicorns

# Of Popes and Unicorns

*Science, Christianity, and*
*How the Conflict Thesis Fooled the World*

DAVID HUTCHINGS AND JAMES C. UNGUREANU

OXFORD
UNIVERSITY PRESS

# OXFORD
## UNIVERSITY PRESS

Oxford University Press is a department of the University of Oxford. It furthers the University's objective of excellence in research, scholarship, and education by publishing worldwide. Oxford is a registered trade mark of Oxford University Press in the UK and certain other countries.

Published in the United States of America by Oxford University Press
198 Madison Avenue, New York, NY 10016, United States of America.

Library of Congress Cataloging-in-Publication Data
Names: Hutchings, David, 1981- author. | Ungureanu, James C., author.
Title: Of popes and unicorns : science, christianity, and how the conflict thesis fooled the world / David Hutchings and James C. Ungureanu.
Description: New York, NY : Oxford University Press, [2022] |
Includes bibliographical references and index.
Identifiers: LCCN 2021017430 (print) | LCCN 2021017431 (ebook) |
ISBN 9780190053093 (hardback) | ISBN 9780190053116 (epub)
Subjects: LCSH: Religion and science—History—19th century. |
Intellectual life—History—19th century. | Draper, John William, 1811–1882.
History of the conflict between religion and science. | White, Andrew Dickson,
1832–1918. History of the warfare of science with theology in
Christendom. | Intellectual life—Religious aspects—Christianity. |
Truthfulness and falsehood.
Classification: LCC BL245 .H88 2021 (print) | LCC BL245 (ebook) |
DDC 201/.65—dc23
LC record available at https://lccn.loc.gov/2021017430
LC ebook record available at https://lccn.loc.gov/2021017431

DOI: 10.1093/oso/9780190053093.001.0001

3 5 7 9 8 6 4

Printed by Integrated Books International, United States of America

*For Chloe, who always wants to see the actual evidence*

*SDG*

# Contents

*Illustrations appear between page 125 and 134.*

# Acknowledgments

Many people have supported me, encouraged me, advised me, and managed not to get fed up with me (I hope!) during the writing of this book. In particular, James has been a joy and a blessing to work with, and he has kept me on the straight and narrow throughout. Danny Ratcliffe's illustrations are excellent. Martin Steel, Danny Byrne, and Angie Edwards have worked through more drafts than it is worth mentioning; their wise words have been a great help. Tom McLeish inspired me to keep going when there were some early setbacks with the project. My parents, Ian and Lynne, have read every paragraph, maintaining a loving faithfulness that they have shown to me my whole life. My precious daughters, Bethany and Chloe, have been a constant reminder of why the truth matters—may they grow up around it and treasure it. And my wife, Emma, is the perfect teammate through it all—thank you, for everything.

D. Hutchings

I'm grateful to David for taking up this project. Scholars have the tendency of only speaking to each other. But David has shown us that we do this at our own peril. As a secondary school physics teacher, he takes the work of scholars and translates it for readers outside of the academy. Working with him on this book has served as a reminder of how important that task really is. The "conflict thesis" remains alive and well because not only have historians of science and religion misunderstood its origins, development, and popularization but we have failed to speak to that larger audience. So I'm grateful not only for the energy and enthusiasm of his writing but more importantly for the empathy and concern he has shown toward our readers.

J. C. Ungureanu

# 1

# Fooling the World

## Kaysing's Case

Bill Kaysing, like 500 million others, was glued to his TV screen. Unlike them, however, he was not being transported to a state of otherworldly wonder by what he was seeing. Instead, he was positively fuming. As far he was concerned, what was happening right now in front of a billion trusting eyes was nothing other than an outright betrayal of humanity. NASA was taking everyone for fools. Unabashed, on live television worldwide, they were faking the moon landings.

And he wasn't going to let them get away with it.

\*\*\*

Seven years later, Kaysing was ready. His grand project—*We Never Went to the Moon*—was finally finished.[1] After copious amounts of research, he had produced what the world really needed: a definitive and watertight account of what had truly transpired in and around July 20, 1969.

Long before that fateful date, he explained, NASA—under pressure from the government, the Russians, and even its own overconfident spin—had realized it was in big, big trouble. Its promise to land astronauts on the lunar surface by the turn of the decade had proved far more difficult to achieve than anyone had originally envisaged. Worried, embarrassed, and trapped in a corner, the big bosses had realized they had no other choice—they were simply going to have to pretend to do it.

It is tempting to assume that *We Never Went* is just paranoid nonsense, but that is far too simplistic a response. Kaysing was no small-town, establishment-hating, tin hat–wearing nutcase ranting at authority from the outside in—no, he had worked for Rocketdyne, the company that built the engines for NASA's Saturn V rockets. He had enjoyed regular firsthand access to sensitive documents about the Apollo missions. He knew many of the engineers personally. And, what's more, he knew how to build a good case.

Firstly, he pointed out, pretty much all of the internal experiments, trials, and simulations had been disastrous: it was not uncommon for the errors to run into the tens of thousands per test. Indeed, he happened to know that the odds of a successful trip had been calculated in-house as just 1 in 60,000—and yet here was NASA, the new darling of the American populace, now claiming to have completed two of them, entirely untroubled, in a row.

Secondly, the photographic evidence on offer was highly suspect. Stars were missing; shadows pointed in the wrong direction; video was unjustifiably grainy; moondust was mysteriously undisturbed. If this was supposed to be proof, it left much to be desired.

Thirdly, Kaysing also noted that the astronauts involved with Apollo only ever seemed to end up in one of two states: rich or dead. In his own words, some arrived back to Earth and were suddenly appointed as "executives in large corporations"; others died before their missions in suspicious "accidents." Could it be that those who played along were rewarded, while those who refused were unceremoniously bumped off?

His arguments went on: diplomatically, the United States needed to overtake the Soviets in the space race; the previously unclassified Apollo documentation had been suddenly rendered "unavailable to the public"; NASA had covered up multiple rocketeering disasters in the past; the moon footage bore more than a passing resemblance to *2001: A Space Odyssey*, which had been filmed during the same period.

Busily quoting panicking engineers, skeptical statisticians, and wary historians, Kaysing painted a picture of failure, desperation, and conspiracy which was hard to ignore. Others soon joined him, citing evidence of their own. Some physicists believed the high levels of space radiation would have finished the crew off long before they had completed their mission. Innovative sleuths unearthed photos of the astronauts—supposedly on the moon at the time—not even wearing their spacesuits. The doubts grew. Would NASA eventually cave? Was it time to admit the game was up?

The rumors continued to rumble and, in 2001, Fox TV decided to press the national body hard on the issue. The documentary *Conspiracy Theory: Did We Land on the Moon?* featured interviews with NASA spokespeople, investigative journalists, astronauts, scientists, and even Kaysing himself. Its conclusion? The question, it decided, remained genuinely open.[2]

## The Truth About Lies

Fox's show and its non-verdict are now two decades old—so what has happened since? Has the explosion of internet use, better home technology, more public access to government files, and the overwhelming force of social media led to the exposure of a corrupt space administration? Should Neil Armstrong go down in history as a gifted actor rather than a swashbuckling pioneer? Had Kaysing spotted the truth before anyone else managed to?

Well, the simple fact is this: the Apollo 11 mission really did succeed in putting men on the moon. Kaysing was wrong. Since his death in 2005, new and far clearer photographs have been taken of the landing area. Sure enough, the telltale signs are there: astronaut footpaths, abandoned vehicles, assorted debris—and a quick online search can confirm it.[3]

None of this, by the way, would come as much of a surprise to David Grimes. During a period of research at Oxford University, the imaginative physicist applied his mathematical wherewithal to the analysis of multiple historical conspiracies. His exhaustive study confirmed what we might already suspect: the higher the number of people asked to keep a secret, the smaller the chance of it actually being kept. Putting it more bluntly, large-scale attempted cover-ups don't work.

The implications for Kaysing's theory are not great: NASA employed no fewer than 400,000 staff during the Apollo program. Grimes's model, applied on this scale, tells us that the likelihood of keeping them all quiet for any longer than five years isn't just low—it is zero.[4]

Interestingly, formal photographs and expert research haven't proved enough to convince everyone—an estimated 20 million Americans still believe that NASA really did make the whole thing up.[5] The new images, they maintain, are also fake. And, why, they might ask, should we believe an Oxford-based finding? After all, they are probably in on it anyway.

Before we despair too much, though, we should remember that most people are not 'hoaxers'—there are, thankfully, more than 300 million US citizens who readily accept the reality of NASA's feat. Most people, even in this high-profile example, have not been duped. Even when one has a semi-convincing and famous case, it seems, the majority of the public won't buy it. The man and woman on the street are a lot harder to fool than many would have us believe.

Of course, if tricking average Janes and Joes is tough, then getting one over on a group of well-informed professionals is nigh-on impossible. No scientist of any repute, for example, thinks the moon landings are fake (Kaysing, as it happens, was a jobbing technical writer at Rocketdyne, and had no scientific qualifications). Similarly, there are no ranking engineers who think the World Trade Center attacks were an inside job; there are no elite historians who think the pyramids were built by aliens.

In short, the people who are paid to know their stuff usually do—or, at least, they know it well enough not to be fooled by conspiracy theorists. And, sooner or later, their expert views trickle down to the public—which kills the story (for the vast majority, anyway) stone cold dead.

So here's the thing: it is a very difficult thing to con the world. Getting even a handful of people to believe in a grand, all-encompassing untruth is hard. Winning over greater numbers than that is extremely challenging. Deceiving just one expert, even for a short while, is highly improbable. The idea, then, that someone might get a significant proportion of the academic community to believe and then promote a lie for a generation or more is utterly beyond the pale.

And yet—staggeringly, shockingly, astonishingly—it has already been done. Nearly 150 years ago, two men who pretty much no one has now heard of set out to convince both the general and the highly educated public of a falsehood—and pulled it off. They have fooled the minority, and the majority. They have fooled a lot of the experts. Their alternative version of history—one which is quite easy to show is untrue—remains the most common view to this day. Somehow, against all of the odds, they have successfully fooled the world.

So, who were they? What did they say? And how, exactly, did they get away with it?

## The Last of the Polymaths

Occasionally, as the years roll ever onward, certain types of people simply vanish. Blown aside by the winds of change, they are suddenly no longer needed—their roles become obsolete, and they are quietly transformed into distant memories of a past age. Think highwaymen, milkmaids, alchemists, or knights in shining armor—once commonplace, they have been swallowed up by time. Each has gone the same way, lost to all but stories.

In our own day, the phenomenon of subject specialism has added yet another victim to this list: the polymath, or master of all trades. In the university of today, the expectation—or even the idea—that someone might make a major contribution in more than one discipline is simply not there anymore. Polymaths are dead and gone.

When modern scientists win a Nobel Prize, for example, their achievement is usually so focused and intricate that even highly informed correspondents struggle to explain it. In a climate like this—one where advances tend to come in minute technical detail, and are understood by so few—it is easy to see why we might have run out of polymaths. After all, who could possibly have the time or ability to excel in more than one field? Wide-ranging, multi-faceted thinking has been consigned, it would seem, to the past.

There is something remarkably sad about this. Modernity is missing out by constraining genius with such tight one-subject-only cords. We will have no Zhang Heng; we will have no Avicenna. Zhang, in the second century AD, wrote entirely new forms of poetry, compared histories of the world to find errors, calculated pi, and invented a machine which could measure both the size and location of earthquakes. Avicenna opened up the second millennium by writing proofs of God's existence, devising thought experiments about consciousness, showing that light must have a finite speed, perfecting steam distillation, and arguing—against Aristotle, no less—that the stars produced their own light.

Specialism has weeded out the polymath from the gardens of today—so, to find the last of them (or, at least, to find them in any significant numbers), we must go back in time. Let us head then, to AD 1811—and to a small town found somewhere in the north of England.

John William Draper was born in St. Helens that year. The son of an itinerant Wesleyan minister, he spent his youth being dragged around by his father's varying employment from church to new church. Worried about the unsettling nature of all this, the clerical body asked for a school be founded in the north, and eventually Woodhouse Grove was established. Young Draper was promptly packed off there aged 11, and lasted a few years before he came home again. He then found his way to university in London, where he studied chemistry. It was at that point that everything changed for him: his father died.

This shook the whole family, and resulted in a dramatic move: Draper, along with his mother and sister, uprooted themselves and headed to Virginia in the United States of America. It was to be here, in the land of promise, that Draper would truly find himself. A man of inquisitive mind and determined character, he threw himself at the fresh opportunities that came his way, always with great success. By the time of his death in 1882, he would be considered by many as worthy of standing alongside the likes of Zhang and Avicenna—for he had become a genuine polymath.

Draper began his path to legendary status with a degree in medicine from the University of Pennsylvania. His work there on the movement of gases and liquids through membranes proved so impressive that he was awarded a professorship in Virginia (1836) and, swiftly, another in New York in 1839.

The very next year, he made technological history: he took the first ever clear photographs of a human face, and then of the moon. Changing tack once more, he went on to demonstrate that metals of different types would all begin to glow at the same temperature (the Draper point); he matched flame colors to the substance being burned; he unearthed a fundamental link between a radiating material's multi-colored "spectral lines" and its composition.

He didn't stop there, either. Soon, Draper discovered that certain combinations of objects could be "electrified" if pulled apart, and used this finding to improve the design of batteries. He helped with the development of the telegraph, making it economically much more viable. He made huge steps in the way we think about physiology by attributing the processes in our body to the laws of chemistry and physics, and rejecting the popular (and vague) idea of "living force."

Such was Draper's brilliance that even the vast and burgeoning realm of science was not enough to contain him. Branching out yet again, he joyfully immersed himself in a hotchpotch mixture of history, philosophy, politics, sociology, and religion. Re-energized, he began to write down some of his own thoughts on all these—and, to his great pleasure, his writings proved popular. Witty, insightful, and ridiculously wide in scope, his prose had a rhythm and drive that raised it above the usual esoteric ramblings of other enthusiastic academics—he was quite the hit.

In 1863, he published *A History of the Intellectual Development of Europe*. Like a master storyteller, Draper drew together the tales of myriad countries, centuries, governments, wars, cultures, religions, and scientific theories to form a single overarching narrative of relentless, positive, human progress—progress which, he believed, was driven by some deeply mysterious and all-encompassing natural law.

It was remarkably well received. The *Westminster Review*, a cutting-edge and radical publication, called it "a noble and even magnificent attempt to frame the induction from all the recorded phenomena of European, Asiatic, and North African History"[6] and went on to say that "Dr Draper soars to a height of eloquence not commonly met with in, yet by no means impairing the cogency of, a strictly philosophic treatise."[7]

It sold well, too. This encouraged Draper to keep writing, and soon came his *History of the American Civil War*, which also became a classic.

By the 1870s, then, Draper's reputation was sky-high. It seemed he could do no wrong—everything he touched turned into intellectual gold. Not only was he considered one of the greatest living scientists in a legendary era of science, but he had become an acclaimed historian, philosopher, and best-selling author to boot. Whenever Draper spoke—no matter the topic—the world, in awe of him, listened.

And, before the end of the decade, he would have much of this listening world hopelessly fooled.

## Conflict

Buoyed by his various successes, Draper now turned his mind to an entirely new topic altogether. What, he wondered, was the real relationship between religious faith and scientific knowledge? Unafraid of controversy, and believing himself to be best placed to comment, Draper decided to write what he hoped would soon be the go-to text on the matter. The title of his study betrayed its conclusion: *A History of the Conflict Between Religion and Science.*

Essentially, Draper's 1874 manifesto ended up describing a war. In its opening salvo, he sets the tone for the rest of the work:

> The antagonism we thus witness between Religion and Science is the continuation of a struggle that commenced when Christianity began to attain political power.

And, rather than remaining a neutral observer, Draper appears, at least, to favor one side of this struggle over the other:

> The history of Science is not a mere record of isolated discoveries; it is a narrative of the conflict of two contending powers, the expansive force of the human intellect on one side, and the compression arising from traditionary faith and human interests on the other.

Science, here, is coming across in a much better light than "traditionary faith." That trend continues:

> As to Science . . . she has never subjected any one to mental torment, physical torture, least of all to death, for the purpose of upholding or promoting her ideas. She presents herself unstained by cruelties and crimes. But in the Vatican—we have only to recall the Inquisition—the hands that are now raised in appeals to the Most Merciful are crimsoned. They have been steeped in blood!

*Conflict*, then, is uncompromising in its critique of organized religion, and gushing in its praise of freethinking science. Bishops are its baddies; geologists its goodies. Interestingly, and despite his seemingly firm decision to opt for one team over the other, Draper maintains that he has always "endeavoured to stand aloof, and relate with impartiality their actions."

The chief finding of his "impartiality" is this: religion will always fight against and try to hold back science, to the great loss of everyone everywhere. Draper—the polymath son of a minister—sums up his case with a startlingly stark ultimatum:

> Then has it in truth come to this, that Roman Christianity and Science are recognized by their respective adherents as being absolutely incompatible; they cannot exist together; one must yield to the other; mankind must make its choice—*it cannot have both*.[8]

Draper, then, had produced a decisive account for the masses. *Conflict* spanned more than two millennia of people, and ideas, and events. It was a wise and learned tome, one eminently worthy of ending any debate once and for all. In his own words, "No one has hitherto treated the subject from this point of view." And that, he felt, was that.

Yet despite his popularity, and experience, and breadth of knowledge, and clout, Draper's final conclusion was still a controversial one. Perhaps, on its own, *Conflict* might not have been quite enough to settle the matter.

The thing is, though, it wasn't on its own—not for long.

## A Brave Man, Wronged

Andrew Dickson White (1832–1918) was born in New York with a family background in both money and education. Initially, his parents expected him to go into the ministry, but he found the first steps on that journey both boring and unpleasant, and ran away from his clerical schooling. Eventually, his father relented and let him study history and English literature at Yale instead.

Once there, White quickly showed his brilliance. He won a number of prizes for his essays and public speaking, including one valued at $100—the highest award available at any university in the world at the time. After graduating, he traveled around Europe for three years and, upon his return, took up a professorship in history and English literature in Michigan.

Lecturing away to his heart's content, White was free to cover an incredible array of subjects over the next few years. He taught on the Roman Empire; the rise of cities; the Crusades; the growth of papal power; medieval Christianity; Islam; parliament in England and France; the revival of learning and art; Luther; the Jesuits; the Thirty Years War; Louis XIV, XV, and XVI; the French Revolution, and even the history of philosophy.

During his European excursions and his subsequent time in Michigan, a dream had begun to well up in White's heart. With each day that passed he longed, ever more deeply, for his very own university. His, he had decided, would be an institution free from any kind of dogmatic oversight; a setting in which students could be genuine and unshackled freethinkers. White believed that, in an environment like this, their uncontrolled minds would settle happily upon such virtues as goodness, purity, selflessness, and peace. This vision moved him profoundly—but how, he wondered, could it ever actually come about?

Then, in the 1860s, White's rich lecturing career was brutally interrupted by the American Civil War. He was promptly nominated for and elected to the New York State Senate (which was a bit of a surprise to him), forcing him to resign from his beloved teaching and research. It was not all bad news, though—for, in this new and unwanted role, White finally stumbled across the chance to fulfill his personal destiny.

Hope had appeared in the form of White's new fellow senator, Ezra Cornell (1807–1874). Cornell was seriously rich—but, as a committed Quaker, he was not all that interested in hoarding cash, or buying luxuries, or collecting trinkets. Instead, he wanted to do something worthwhile. White, of course, was ready with a suggestion: together, the two of them should found an innovative, modern university, one which he promised his potential partner would help contribute to humanity's next big step forward.

Before too much longer, Cornell University became a serious proposal: a bold establishment where religion and dogma would not hold back study and where students would be wholly unfettered. White, the dreamer, was hugely excited. What happened next, therefore, not only shocked him—no, it came close to breaking his heart.

White, for what it's worth, had been extremely careful when explaining Cornell's guiding principles to its trustees in 1866—he clearly felt that he was guiding the world toward the light:

> We have under our charter no right to favour any sect or promote any creed. No one can be accepted or rejected as a trustee, professor or student, because of any opinions or theories which he may or may not hold. . . . Development under this principle—moral, intellectual and physical—can only be normal and healthful in an atmosphere of love of truth, beauty and goodness.[9]

Imagine his surprise and dismay, then, when others began to speak against his project. For, rather than being praised from the rafters, White's enterprising brainchild faced stern opposition from the off.

Much of the criticism took the same basic form: that Cornell would be irreligious. It would open up young minds to dangerous ideas, he was warned—it would lead them astray from Christian truth. If this new place of "learning" was allowed to flourish, his detractors said, it would merely be the thin end of a wedge. Students would be misled, and alienated from God—with disastrous consequences.

White was horrified. He wrote to Cornell, grieving that "the papers, addresses, and sermons on our unchristian character are venomous."[10] Wobbled, but not unseated, the duo pressed on regardless. As a result, the university did indeed open in New York in 1868—and it took in the biggest entering class of any American school in history. It quickly became a very successful operation, and it still is today. Its current motto is taken from Ezra Cornell's opening address that very year:

> I would found an institution where any person can find instruction in any study.

White's anger, however, continued to simmer. For him, the vociferous attacks on his plans had confirmed what he had suspected for quite some time—that dogmatic religion was only ever a dreadful thing. Doctrinal inflexibility of the type displayed by the supposedly holy opposers of Cornell was, he eventually concluded, the mortal enemy of "truth, beauty, and goodness."

And perhaps, he thought, it was time to fight back.

## Warfare

White became a man with a plan: he would bring dogmatism down, and he would use his two biggest guns—public speaking and persuasive writing—to do so. It wasn't long until the first shooting opportunity came along: in 1869, White was invited to give a talk of his choice at Cooper Hall in New York. He accepted—and his resulting speech would forever change our history.

Many in the audience were still reeling from the devastating effects of the American Civil War—families bereft, memories scarred, relationships strained. Despite the likely sensitivities in the room, however, White quite deliberately picked war as his dominant theme—he called his lecture "The Battle-Fields of Science."

Already close to the bone with his choice of motif, White pressed the knife in still further. In his introduction, he told his hearers that he would be describing hostilities "with battles fiercer, with sieges more persistent, with strategy more vigorous than in any of the comparatively petty warfares of Alexander, or Caesar, or Napoleon."[11]

With fire in his belly and Cornell in his head, White began to regale his listeners with story after story of conflict between two violently opposed forces—one heroic, one villainous. The lesson to be learned from history, he said, was a simple one: religion was the enemy of science. He powered on, naming case after case in which dogma had damaged cosmology, or astronomy, or chemistry, or anatomy. It was nothing less than a broadside. And it was nothing more than the beginning.

The very next day, the *New York Daily Tribune*—whose editor, rather conveniently, was a trustee of Cornell—printed the lecture in full. Tongues were set wagging, both in academia and on the street. White was asked to repeat the lecture again and again, all over the country. He did so, adding new explosive material each time. Eventually, in 1876—less than two years after Draper's *Conflict*—White's expanded and amended lecture was published as a pamphlet, for anyone to read at their leisure.

White, however, was not done. He continued to write on the subject, and had his work serialized, for more than a decade, in the well-known and daringly modern *Popular Science Monthly* magazine. Finally, in 1896, his vast survey of history, philosophy, physics, theology, biblical criticism, biology, sociology, and more was pulled together into one totemic volume—a lengthy, exhaustive, and ruthless attack on dogmatic religiosity everywhere. He called it *A History of the Warfare of Science with Theology in Christendom*.

In its introduction, White linked the development of his book directly to his experience with Cornell:

As honored clergymen solemnly warned their flocks first against the "atheism," then against the "infidelity," and finally against the "indifferentism" of the university, as devoted pastors endeavored to dissuade young men from matriculation, I took the defensive.

He had initially reacted with "sweet reasonableness," he said—but that had done "nothing to ward off the attack." Then, finding himself frustrated, upset, and at a loss with what to do about it all, he admits that something just suddenly snapped deep within:

> Then it was that there was borne in upon me a sense of the real difficulty—the antagonism between the theological and scientific view of the universe and of education in relation to it.[12]

This book, then, was his solution: White would change the way people thought about science and religion forever. It was time for the truth to be told, and for the world to move forward. Science—and the freethinking associated with it—must be unleashed. Religious dogma, its age-old enemy, must be finally and decisively killed off. Nothing less than outright victory would suffice.

## The Ultimate Double Act

White's titanic *Warfare* dwarfed Draper's popular-length *Conflict*. It read differently, too. While Draper's book was pithy, witty, and even conversational at times, White's was more like a colossal encyclopedia entry—albeit with the odd outbreak of fury and despair at the sheer ridiculousness of theologians. Each covered roughly the same ground, but White did so in far more detail, and added extra topics on top.

The highly complementary nature of the two works turned out to be remarkably important. Perhaps each book on its own would have been forgotten—but instead, as a double act, they more than reinforced each other. That *Conflict* and *Warfare* appeared in the same era, had the same broad scope, and agreed on so much added a great deal of weight to their claims. Even their dissimilarities were beneficial—they appealed to different readers, and so their combined message spread both far and wide.

Consider the following, for example: Draper took aim at the Catholic Church; White held that Protestants were just as bad, if not worse. Draper never revised his book once; White's was the constantly tweaked accumulation of more than twenty years of essays. Draper wrote as a world-renowned scientist; White was a famously gifted politician-historian. *Conflict* included not a single footnote, making for a smoother read; *Warfare* had literally thousands of them, suggesting airtight reliability.

In short, they formed an astonishingly effective pincer movement. When both books were considered together, their case appeared to be fully made.

Such was the combined influence of *Conflict* and *Warfare* that they genuinely managed to achieve precisely what we earlier stated was impossible. They sat on bestseller lists for decades, sold all across the globe, were translated into all sorts of languages, and became the twin final words on their topic.

Between them, then, John William Draper and Andrew Dickson White did more than launch a small-scale conspiracy theory, or gather a handful of limited but loyal followers for a couple of years. Instead, they fooled the world.

## Making Myths

A major point in both texts was this: it is a pitiful and embarrassing thing for grown and intelligent people to believe in myths. The Bible's stories of miracles, they claimed, were little better than fairy tales—science, if only people would listen to it, could free them from such imaginary nonsense. And yet, in their desperation to dismiss dogma, our two authors managed to fall foul of their own accusations—for, during their two projects, they uncritically accepted, wholeheartedly believed, and then willfully propagated an entire collection of myths of their own.

For example, *Conflict* and *Warfare* portrayed a Christianity that killed off Greek and Roman science and plunged Europe into the so-called Dark Ages—a thousand years of fecklessness, stupor, and backward thinking. Christendom, said Draper and White, destroyed ancient libraries; it forbade the study of philosophy; it tormented or tortured or killed anyone remotely scientific lest its hopelessly inaccurate scriptures be exposed.

There was more: the ever-dogmatic Church, they explained, maintained that the Earth was flat (until it was proved wrong by Columbus); it tried to strangle science at the first signs of the Enlightenment; it resolutely rejected any concept of natural law. It hated Nicolaus Copernicus for dethroning the Earth; it decried all claims that the Earth moved; it hurled the heroic Galileo Galilei into prison; it burned Giordano Bruno for his astronomical foresight.

Worse still, institutional Christianity also blamed all illness on demons; it prohibited medical intervention of any kind; it banned dissection; it banned autopsies. It refused anesthetic to women in labor. Interpreters of Holy Writ had unanimously denied heliocentrism, evolution, an old Earth, and a vast universe until their puny sacred arguments could no longer bear the weight that had been placed on them. In short, the Church had stood, with moronic steadfastness, against all scientific endeavor for more than 1500 years—until, as a long-overdue reward for their painful perseverance, the brave and logical scientists finally broke free.

All these grand and saddening assertions have three things in common. Firstly, they form the backbone of the Draper–White narrative. Secondly, they have since become common knowledge, and are repeated in casual conversation, newspaper articles, popular science and history, plays, newspapers, documentaries, and even academic treatises. Thirdly, we now know that none of them—not a single one—is actually true.

## A Legacy of Lies

The series of myths that Draper and White spread about science and religion are known today in the literature as the *conflict thesis*. Thanks to the dedicated and committed research of a band of specialists operating since the 1980s at least, the conflict thesis has now been thoroughly debunked. One by one, the tales spun out in *Conflict* and *Warfare* have been shown to be either entirely false, horribly misunderstood, or deliberately misrepresented.

Ronald L. Numbers, for example, is perhaps the foremost historian of science and religion alive today. Here is his damning assessment of the conflict thesis:

> The greatest myth in the history of science and religion holds that they have been in a state of constant conflict. No one bears more responsibility for promoting this notion than two nineteenth-century American polemicists: Andrew Dickson White and John William Draper. . . . Historians of science have known for years that White's and Draper's accounts are more propaganda than history. . . . Yet the message has rarely escaped the ivory tower. The secular public *knows* that organized religion has always opposed scientific progress. . . . The religious public *knows* that science has taken the leading role in corroding faith.[13]

This passage is from the introduction to *Galileo Goes to Jail*—an edited volume, published by Harvard University Press, which consists of essays by twenty-five of the world's very best thinkers on these issues. There is a clear, evidence-based consensus among this group: the conflict thesis is utter bunk. Yet the frustration, on Numbers's and his colleagues' part, is clear from the excerpt: no one, it seems, is actually listening to them.

Over the last century and a half, despite its untruth, the God-or-science mantra has become firmly embedded in our culture. Indeed, Draper's forceful summation of it—his "cannot have both" formulation—can be found almost everywhere. A 2013 survey of high school students in the United Kingdom, for instance, found that the majority agreed with the statement "the scientific view is that God does not exist."[14]

These youngsters are not alone. The New Atheist writers—the likes of Richard Dawkins and Sam Harris—have been rehashing Draper and White, in one way or another, for years. As an avid polemicist, Dawkins is perhaps the duo's most famous intellectual descendent—indeed, he struggles to get through even a paragraph about science before reminding his readers that "Religious beliefs are dumb and dumber: superdumb."[15]

In the meantime, Dan Brown—bestselling author of more than 250 million novels—has done even better than Dawkins and made far more than just a name for himself from Draper's and White's ideas. In *Angels and Demons* (think a heroic particle physicist and a morally questionable pope) the legacy of both *Conflict* and *Warfare* is palpable:

> "Mr. Langdon, all questions were once spiritual. Since the beginning of time spirituality and religion have been called on to fill in the gaps that science did not understand. The rising and setting of the sun was once attributed to Helios and a flaming chariot. Earthquakes and tidal waves were the wrath of Poseidon. Science has now proven those gods to be false idols. Soon *all* Gods will be proven to be false idols. Science has now provided answers to almost every question man can ask. There are only a few questions left and they are the esoteric ones. Where do we come from? What are we doing here? What is the meaning of life and the universe?"
>
> Langdon was amazed. "And these are questions CERN are trying to answer?"
>
> "Correction. These are questions we are answering."[16]

Should we really take Brown or the New Atheists as evidence, though, that Draper and White have got to everyone? After all, Brown is writing make-believe, and the New Atheists are hocking just about anything that opposes religion. They can hardly count as "everyone," then, can they?

Peter Byrne, on the other hand, is a different case altogether. As an award-winning investigative reporter and a science writer of some repute, Byrne has no need to invent storylines (as per Brown), and has no obvious axe to grind (as per Dawkins). In one of his books, however—an acclaimed biography of the quantum theorist Hugh Everett—we can find the following: "Vatican attacks on a scientific theory usually are a sign that it is intelligent and correct—just ask Copernicus, Galileo, Bruno, Newton, Darwin, and Einstein."[17]

Here, then, in a book which is not even interested in their subject matter, the spirit of *Conflict* and *Warfare* remains both alive and well. Byrne's assertion is straight out of the Draper–White tradition—indeed, the pair would probably have been proud of such a line.

It is a good line, too—but it is wrong. Copernicus himself was ordained, and was never opposed by any church.[18] Galileo could not prove his heliocentric case scientifically, but half of the Vatican supported him regardless of that fact.[19] Bruno got into trouble for his theology, not for his science (which was, as it happens, not really the type of science that Byrne might have imagined, anyway).[20]

The Catholics liked Newton's theories, for he explicitly wrote that they came from, and required, God. Darwin's work was admired by just as many believers as despised it—the Catholic Church, incidentally, favors evolution.[21] Einstein's deductions faced no religious opposition at all. Instead, his ideas were embraced by the Catholic priest Georges Lemaitre, who then used them to devise the Big Bang model of the universe.

Byrne, then, is mistaken on all counts—so how did he ever come to write such a sentence? He is, quite clearly, not stupid. His text is not the self-published rant of an angry conspiracy theorist. His publisher is Oxford University Press. His title is endorsed, among others, by *New Scientist*, by the BBC, and by a Nobel Prize–winning physicist. What is going on here?

Well, Byrne's book is not actually about the conflict thesis. Neither is it about science and religion. Instead, it is a biography of a quantum physicist, and the statement we picked out is in no way central to its theme. If anything, it is an isolated and throwaway line. There is no footnote. The topic is not dwelt on any further. It is, at best, an aside.

Strangely, though, the fact that it is an aside makes it more important, not less. By simply nodding at the conflict thesis like this, Byrne implicitly champions it. He assumes it to be common knowledge; he feels that no further explanation is necessary. There are no two ways about it: he has been duped. Somehow, reaching out across more than a century, Draper and White have got to Byrne—who, we must remember, is an expert in comparison to most—just like they have got to almost everyone else.

Still, does it really matter all that much?

## Harmful in Itself

Why should anyone outside of a lecture hall care in the least about the conflict thesis? Can't this whole debate quietly play out in the libraries of our universities, just like many other idiosyncratic and irrelevant disagreements that secure the steady study of scholars on second-rate salaries?

The answer is no. David Aaronovitch, author of the myth-busting *Voodoo Histories*, explains why:

> There is a more sinister aspect to jovial arguments about whether or not the moon landings actually took place, and to the speculation about why we enjoy such arguments. The belief in conspiracy theories is, I hope to show, harmful in itself. It distorts our view of history and therefore of the present, and—if widespread enough—leads to disastrous decisions.[22]

Disastrous decisions? Isn't Aaronovitch being a little over-dramatic here? Well, perhaps we should ask Michael Reiss—an evolutionary biologist who, in 2008, lost his job with the Royal Society after a journalist misunderstood his comments on creationism and the modern *Conflict-Warfare* gang got hold of him. Sir Harry Kroto, a Nobel laureate, demanded Reiss's head immediately. Channeling his inner Draper and White, Kroto wrote, "There is no way that an ordained minister—for whom unverified dogma must represent a major, if not the major, pillar in their lives—can present free-thinking, doubt-based scientific philosophy honestly."[23]

Or ask Elaine Howard Ecklund, an acclaimed social scientist working at Rice University who, in 2016, surveyed nearly 10,000 scientists around the world on the matter.[24] She found that even though more of them described themselves as religious than as atheists, there was a dominant elitism coming from the smaller group. Discrimination on the grounds of religion was commonplace, she discovered. In many cases, this included bullying, name-calling, discrediting individuals or groups, and even passing them over for jobs.[25]

Or ask Tom McLeish, decorated physicist and fellow of the Royal Society who, in 2017, dared to write a book about his Christian faith entitled *Let There Be Science: Why God Loves Science, and Science Needs God*. Influential American biologist Jerry Coyne was having none of it, and was especially enraged by McLeish's significant role in public science. In his review of the book—during which he admits that he has not actually read it—Coyne sneered, "Chair of the Royal Society's education committee? What the bloody hell is a theist doing in *that* position?"[26]

His readers—who number in the tens of thousands—then proceeded to stir themselves into a fury at the thought of a practicing scientist having a faith. One of them even asked Coyne directly, "don't you weary of fighting these assholes?"

*Conflict* and *Warfare*, it would seem, have become dangerously self-fulfilling prophecies—it can get quite horrible out there at times.

Which, by the way, is the last thing that Draper and White would ever have wanted.

## A Twist in the Tale

Draper himself was no atheist; neither was White. They were not even agnostic. In fact, both thought of themselves as followers of Christ, and viewed their books as significant contributions to his cause. They were writing, the two of them said, not to push science and religion ever further apart, but instead to bring them both back together.

Here is White, in the preface of *Warfare*: "My conviction is that Science . . . will go hand in hand with Religion." Indeed, he preaches like a revivalist, extolling the teachings of Jesus:

> Religion, as seen in the recognition of "a Power in the universe, not ourselves, which makes for righteousness," and in the love of God and of our neighbor, will steadily grow stronger and stronger, not only in the American institutions of learning but in the world at large. Thus may the declaration of Micah as to the requirements of Jehovah, the definition by St. James of "pure religion and undefiled," and, above all, the precepts and ideals of the

blessed Founder of Christianity himself, be brought to bear more and more effectively on mankind.[27]

Draper, for his part, also saw the historical God-versus-science battle as unnecessary—in *Conflict*, he called for "a friendship, that misunderstandings have alienated, to be restored."[28]

As Numbers has already explained, though—and as we have already begun to detail here—that is precisely the opposite of what actually happened.

## Of Popes and Unicorns

We are left, then, with quite a few questions. If these two men intended to reconcile God and science, then how did they manage to make such a big mess of it? If they were both smart, well read, and respected, then how did they come to write such error-strewn manuscripts? If their books were indeed packed with fables and misinformation, then how did they ever gain such a firm foothold among the educated upon their release?

We can keep going: why does the conflict thesis hold such a strong grip now, more than a century after *Conflict* and *Warfare* were written? How is it that so many elite academics—think Kroto and Coyne, for instance—continue to endorse their wrongheaded ideas? If they were so famous and successful at the time, then why has hardly anyone today even heard of Draper, or White, or their two books? And, for that matter, what actually is the truth about science and religion across history?

These questions can lead to still more: Why did so many prominent scientists—Kepler, Boyle, Faraday, Maxwell—attribute their scientific advances so directly to their faith? Why are the New Atheists so sure that the two are incompatible? What makes historians of science, like Numbers, say that the conflict thesis is drivel? Do we have to choose between God and science, or not?

This book, then, will be the story of two books. We shall look at their authors, and at the world that they lived in. We shall look at their content, and at where they went wrong. We shall discover their extraordinary influence as we see their dubious claims repeated, again and again, in the work of countless commentators nearer our own time—commentators, by the way, who should really know better. We will correct their false narratives. We will chase down the truth.

This book is the story of flat earths; of dissection, of autopsies, and anesthetic; of creation and evolution; of laser-eyed lizards and infinite worlds; of miracles and of man-made mountains; of gods; of popes and unicorns. It is the story of comets and mathematicians; of souls and libraries; of the Greeks, the Enlightenment, and the Not-So-Dark-After-All Ages. It is the story of Galileo, of hot dates on spacecraft, and of immortal peacocks. It will be a journey through time, through cultures, through ideas, through personalities, and—ultimately—through the human condition.

And, perhaps, it will result in what we are all, ultimately, looking for: no more *Conflict*; no more *Warfare*.

# 2

# Lone Voices?

## Newton and the Hairdresser

Ask any practicing scientist to list their top ten thinkers of all time, and it is likely that Isaac Newton (1643–1727) will make the cut. This is with good reason: he personally formulated three laws of motion which are still taught to every physics student in the world; he painstakingly analyzed the colored spectrum of visible light; he invented the now-essential mathematical calculus; he built the first ever reflecting telescope. And, if all that wasn't already enough, then there's also the small matter of gravity—which he compre-hensively described in his game-changing *Philosophiæ Naturalis Principia Mathematica* (1687).

Maurice Ward, on the other hand, will not get a mention from an-yone. A hairdresser from Hartlepool (a town whose occupants are most famous for convicting a monkey of being a French spy and publicly executing it[1]), Ward would not make their top thousand, let alone trouble the likes of Newton. Born in 1933, he achieved precisely nothing in the formal world of academia, is unheard of in the halls of our greatest uni-versities, and died without a single law, principle, or even journal paper to his name.

And yet, despite the incontestably lopsided pen-pictures above, Ward can still make a genuine claim to be Newton's scientific superior. For, like Newton, Ward came up with an invention that could change the world—but, unlike the much-lauded Enlightenment powerhouse, the barely known hair-dresser did it all, entirely, on his own.

Let's deal with Newton's long list of spectacular achievements first. His laws of motion were indeed paradigm-shifting—and yet he was, in reality, building on the breakthroughs of Galileo Galilei (1564–1642). In the same vein, Newton's immediate predecessors—Willebrord Snellius (1580–1626) and René Descartes (1596–1650), for example—had laid much of the neces-sary groundwork for his ideas about light.

The pattern continues: at almost the same time as Newton devel-oped calculus, Gottfried Leibniz (1646–1716) came up with it too—a fact which strongly suggests that mathematicians had recently created the ideal conditions for the discovery. The reflecting telescope was indeed a clever new design, but it was a modification of pre-existing concepts. Even Newton's gravity was highly dependent on the equations of Johannes Kepler (1571–1630), as well as on suggestions put forward (sometimes to Newton directly) by Robert Hooke (1635–1703).

None of this, of course, should serve to detract from Newton's genius—he is undoubtedly one of the greatest minds in human history. It is true, how-ever, that he lived and worked in an era when other great minds were making highly significant contributions in the same arenas. His community included the likes of the nascent Royal Society, and mathematical and experimental leaps forward were being made left, right, and center. Newton, then, was not a lone voice in the wilderness. He was not an isolated savant. As he famously wrote in a letter to Hooke, he was "standing on the shoulders of giants."[2]

Maurice Ward, however, was a different kettle of fish.

## Selling Starlite

Ward, like much of the UK population, had been deeply affected by the horrific deaths of 55 passengers on British Airtours Flight 28M in 1985. The aircraft, a Boeing 737, had caught fire while still on the Manchester runway, and only 82 of the 137 people on board managed to escape. The tragedy was due mainly to two factors: the great speed with which the fire spread, and the deadly nature of its fumes. The overwhelming majority of fatalities were ascribed to "rapid incapacitation due to the inhalation of the dense toxic/irritant smoke atmosphere within the cabin."[3]

Despite a lack of any notable background in science or technology, Ward set about developing a fire-resistant material that would ensure such a disaster could never happen again. He did not consult with any professionals, he did not book any laboratory slots, and he did not build on the discoveries of others. Instead, he worked at home—with occasional help from his (equally non-qualified) daughter.

Against all the odds, though, Ward developed a miracle material which outperformed anything that has been produced by any multinational engineering company before or since. He christened it *Starlite*—and its list of properties staggered even the experts.

Starlite was tested by the likes of NASA, NATO, the British Ministry of Defence (MOD), Cambridge University, and even Boeing itself. Between them, they found that just a few millimeters of the substance would comfortably withstand laser beams at 10,000 °C, sustained flame temperatures of well over 1000 °C, and even the full and direct heat blast from a nuclear explosion. What's more, it did all this without transmitting any warmth across itself or emitting any dangerous fumes.[4] It was the perfect defense against the threat of fire—and the miracles do not stop there.

Starlite, according to Ward and his daughter, is more than 90% natural. It is fully biodegradable and, they claimed, even edible. It is flexible when dry, can be painted onto a surface when wet, is waterproof, UV-proof, and remarkably light. Such is the adaptability of the product that it could potentially be sprayed harmlessly over lawns or gardens to protect the houses within from the horrors of wildfires.

This wonderstuff hit the big time when it got the prime slot on the BBC's flagship science show *Tomorrow's World* in 1990.[5] Presenter Peter Macann subjected a hen's egg coated in a thin layer of Starlite to five minutes of blowtorching from about an inch away. He then turned the blowtorch off, immediately picked the egg up in his hand (with no ill effects) and cracked it into a dish. To the delight of the viewers (and Macann) it was raw.

The big-hitters, unsurprisingly, promptly piled in. Ward found himself with potential suitors all over the world. He sent samples of Starlite out for them to test, and the verdict was unanimous—this was an unprecedented material that could change engineering forever. The world could be a much safer place. Ward could be a much richer man. Starlite could be a household name.

And yet, it was not to be. Ward, the quirky outsider, simply could not get along with the wider scientific community he was suddenly plunged into. He was fiercely defensive when it came to discussing his formula and methods and he continually resisted collaboration of pretty much any kind. Mark Miodownik is one of the foremost materials scientists in the world, and has personally tested Starlite. When asked why it was never brought to market, he says:

> It's a very different skillset, inventing material, and building a business out of it. Those are, in a sense, two completely separate things and I think he either couldn't do the second thing, or he didn't have someone around him he trusted to do the second thing.[6]

Keith Lewis, who performed the laser tests for the MOD and described Starlite as better than anything else he had ever seen, agrees that Ward's stark individualism was his downfall:

> I think Maurice was very, very bothered about knowledge leaking out. He wanted to own it. He wanted Starlite to be "his." He was not happy with giving over details of what might be in it.[7]

As the years rolled on, Maurice Ward proved more and more difficult to work with. Frustrated, the big companies began to lose interest. Eventually, they cut their losses and moved on. Ward never did sell Starlite. And then, in 2011, he died—taking his secret with him.

The miracle material that could have revolutionized our millennia-old relationship with fire was lost forever. Many well-funded and well-educated experimentalists have tried, and failed, to replicate it. The conclusion is as clear as it is surprising: Ward and his invention, it would seem, were both one of a kind. The undecorated and previously unheard of hairdresser from Hartlepool was clearly not standing on anyone's shoulders—let alone on those of giants.

## Lone Voices

So what about John William Draper and Andrew Dickson White: were they Newtons, or were they Wards? Were they talented and respected members of an active and interconnected academic community, or highly individualistic lone wolves? Did their grand masterworks—*Conflict* and *Warfare*—emerge brilliantly but naturally from the thinking of the day, as with *Principia*? Or, more like Starlite, did they suddenly appear, without any precedent whatsoever, from the minds of disconnected and idiosyncratic outsiders?

Essentially, what we really want to know is this: is the conflict thesis entirely of our duo's own making, or did they have (lots of) help?

In May 2015, a group of the world's top historians of science and religion gathered together for a three-day conference at the University of Wisconsin–Madison. Messrs. Draper and White, as one might have expected, came up more than just once. So, what was made of our prickly pair? Were they deemed to be lone voices, or part of the chorus?

Well, as is often the way with such events, the various discussions ended up in a book. Published in 2018 by Johns Hopkins University Press, it was given the more-than-a-little-revealing title of *The Warfare Between Science & Religion: The Idea That Wouldn't Die*.

No fewer than ten of the seventeen essays in this wide-ranging volume—chapter titles include "Continental Europe," "Social Scientists," "Muslims," and "The View on the Street"—mention Draper and/or White on their very first page. A further two introduce them just one page later. That is a rather considerable hit rate of twelve from seventeen—it would seem, then, that the specialists simply cannot separate the notion of God-and-science enmity from our two pugnacious penmen.

Lawrence M. Principe, for example, is professor of the humanities at Johns Hopkins. He is also the director of the Charles Singleton Center for the Study of Premodern Europe, and has been awarded the Francis Bacon Medal for significant contributions to the history of science. In his section of the book, he could hardly be accused of holding back:

> Historians identify two late-nineteenth-century books as the chief vectors of the conflict thesis: John William Draper's *History of the Conflict between Religion and Science* (1874) and Andrew Dickson White's *A History of the Warfare of Science with Theology in Christendom* (1896). The melodramatized "history" and the dubious "facts" upon which these books rest are easy to point out and refute, and many historians have done so.[8]

Elsewhere, Bernard Lightman says Draper and White are "generally seen by scholars as the founders of the 'conflict thesis'"; Ronald L. Numbers and Jeff Hardin call them "the chief architects of the 'warfare thesis'"; M. Alper Yalçinkaya says they "introduced" the God-versus-science idea; Efthymios Nicolaidis names *Conflict* and *Warfare* as "the two most influential nineteenth-century books positing the existence of 'warfare' between science and religion."[9]

This is pretty damning. Does all this serve to tell us, then, that the two of them really are solely responsible—as in, Maurice Ward–style solely responsible—for the existence of the badly-wrong-but-widely-accepted conflict thesis? Is it exclusively their fault that so many school students now believe we must choose between Science or God as our Guide? Does the whole darn show hang on them—and on them alone?

Well yes, sort of.

And also no.

## Partners in Crime?

When we dig a little deeper into the articles written by the Wisconsin collective, new structures begin to emerge. Principe, for example, hints that Draper and White might not have been the first to attempt their particular summit: "A few limited claims that 'religion' (broadly understood) opposed 'science' (equally broadly understood) do predate the nineteenth century."[10]

Jon H. Roberts, professor of history at Boston University, makes a different but related point—he says of those who first reviewed Draper's *Conflict* that:

Many acknowledged that the relationship between religion and science was, as one commentator put it, "*the* question which is now agitating the world of thought." Many recognized, too, that Draper was hardly alone in perceiving significant tension between religious thought and the scientific world view.[11]

If, though, it is true that some others had mentioned, fomented, or even propagated the conflict thesis before or alongside Draper and White, then who were they? What, exactly, had they said? Were Draper and White aware of them? More to the point, if the conflict thesis was doing the rounds, then does that mean that people were already telling lies, and spreading myths, and getting history wrong? How could such a thing possibly happen? And why would Draper and White end up, more than a hundred years later, getting all the blame for it?

Let's find out.

## Plotting Our Course

Before we continue, a caveat: tracking any intellectual idea back to its earliest origins can be a bit of a fool's errand. Invariably, what starts off as a seemingly simple quest soon becomes an odyssey, often of appropriately epic proportions. What initially appears to be the first seed of a movement turns out, instead, to be a bloom in its own right—one with a still-earlier seed of its own. "First" versions of arguments can have the rather annoying habit of wriggling around, slipping our grasp, and running away. And they tend to run backward, through time.

It is very hard, then, for a historian to ever start a story—for the case can always be made for going just one more step further into the past. And, given that our particular topic is the relationship between science and religion, the risk of indecision is especially bad—for both appear to have been with us, in one form or another, for the entirety of recorded history.

Still, a choice must be made, or we will get nowhere. Here, then, is our itinerary: we shall begin in France, at the end of the 1780s. We shall drop in on some revolutionaries and then, a few decades later, on a troubled philosopher. From there, we shall head to Victorian Britain and visit some high-society freethinkers. Finally, we shall set sail across the pond to America—where, armed with what we have learned, we shall at last find the answers we need about Draper and White.

So, off to France we go . . .

## Two Cults

Owing to their discontent with the economy, or the ruling classes, or the class system itself, or the questionable behavior of those in Catholic orders, or the lavish hypocrisy of the royal family, or their own (lack of) diet—or some combination of all of these—a whole lot of people living in France toward the close of the eighteenth century decided that they had now had enough. A confusing and messy political revolution was embarked upon—one which is almost impossible to follow closely and accurately, even with the benefit of more than two centuries of hindsight.

It is hardly the business of this book to analyze the French Revolution in any grand sense. Instead, we will attempt to pick out just a few key elements from the notoriously complex whole—for both science and religion featured, in their own way, rather heavily.

The revolutionary leaders, for example, despised the Catholic Church and wanted to deprive it of its power. The problem was that they could not agree on precisely how this should be done. Indeed, such disagreement among the *sans culottes*, as they were sometimes known, was a constant feature of the revolution, and one that left many of them dead—often at the hands of each other.

Joseph Fouché (1759–1820) was one of many who thought he knew best what to do. To him, it seemed to make sense to dechristianize France altogether. Under his instruction, churches were stripped of their religious symbols, and their assets were sold to fund the revolutionary efforts. Crosses were even removed from cemeteries, with the atheistic Fouché insisting on the inscription "Death Is an Eternal Sleep" being placed over their gates.[12] Public Christian worship was outlawed; priests were now effectively enemies of the state.

Fouché found an ally in Jacques Hébert (1757–1794). The latter realized, though, that getting rid of Christianity altogether might present its own problems—for much of the rural population were deeply committed to it. He set about devising a highly rational replacement for Catholicism, in the hope that it would both appease and enlighten these countryfolk. Soon, he unveiled his brainchild: the excitingly new and refreshingly godless Cult of Reason.

Officially established as the creed of the Revolution in 1793, the Cult of Reason didn't actually last all that long. Its mayfly-like climax came with the "Festival of Reason," celebrated nationwide in November of that year. Running somewhat contrary to its name and intention, it ended up being an inconsistent and haphazard blend of music, partying, sage proclamations of revolutionary philosophy, nods to ancient Greece, grand displays of art, and wanton rejoicing in the newly bought "freedom" of France.

So unclear, in fact, was the actual purpose of the festival that different parts of the country even gave it different names: some went for the Festival of Virtue, some for the Festival of Morals. Many towns decided to have local women portraying "goddesses," with some claiming that their particular divinity (sometimes Liberty, sometimes Reason, sometimes Fame) was the one and only goddess—no matter what the folk down the road had to say.

Other celebrants missed the point of the charade even more badly, and used the occasion to worship an unspecified but almighty creator. All of this confused and chaotic my-town's-god-or-goddess-is-bigger-than-yours business, we must remember, was happening entirely at the behest of an atheistic cult which had been formed with the sole purpose of championing level-headed, anti-superstitious, clear-thinking reason.[13]

To make matters even more complicated, the head honcho of the revolution at this particular time was Maximilien Robespierre (1758–1794)—who did believe in God. He found himself appalled at the rampant godlessness of the Cult of Reason and its frenzied festival, and decided he needed to rescue his embryonic France from it as soon as possible.

Robespierre, however, was in no man's land—for he, most certainly, was not a traditional believer either. His "God" was a vague and undefined entity, a concept he felt was far beyond human understanding. As a result, he set himself a monumental task: he would eradicate both the ancient Catholic Church and its modern atheistic replacement in one fell swoop.

His solution was to copy Hébert, and form yet another cult: this time, it was the Cult of the Supreme Being. Convinced that his new invention was the answer, Robespierre did not hold back—after all, he was currently enjoying a lot of power and influence, and he might as well use it. He, too, would have a festival—the Festival of the Supreme Being—and his would not be the embarrassing free-for-all that Hébert had made such a mess of. Instead, it would be a carefully choreographed and ostentatiously orchestrated triumph.

Paris, Robespierre commanded, was to come to a standstill as his bold new religion was ushered in. Everyone—no matter what they had believed up until this point—would stand to benefit. The discomfort stirred up by the atheism of Hébert would be relieved by the acknowledgment of the (sort of) supernatural; the corruption and irrationality of the Catholicism that had held the old France back could be binned once and for all. He was, therefore, providing his fortunate country with the best of both worlds.

No expense was spared on the day itself, which began with a lavish parade through Paris. Historian David Andress takes up the story:

Robespierre headed the march. . . . He was reportedly radiant with happiness at the proceedings. . . . After the massive choral performance at the Tuileries, he made another set-piece speech on virtue, then applied a flaming torch to a large effigy of Atheism—from it . . . emerged a smaller figure of Wisdom.

The spectacle continued with a heady, dream-like, mishmash of imagery:

A vast mountain . . . had been built from plaster and board, and alongside it a fifty-foot column with the figure of the French people as Hercules. Crowned with a massive tree of liberty, and supplied with a convenient staircase, the mountain was ascended by the deputies in procession, as choirs, and some half-million spectators, boomed out more newly composed hymns. As the noise crashed to a halt, Robespierre descended like Moses himself.[14]

Magisterial indeed.

Within six weeks, though, Robespierre was dead. Earlier that year he had sent his fellow revolutionary, Hébert, to the guillotine; now it was his turn to be executed by his supposed teammates. The Royalist Protestant Jacques Mallet du Pan (1749–1800) summed the murderous situation up fairly well in 1793: "Like Saturn, the Revolution devours its children."[15]

A few years later, Napoleon became emperor of France. He outlawed both the Cult of Reason and the Cult of the Supreme Being in 1801 and, in the process, restored Catholicism's civil status in the country. The short and exceptionally bloody revolution was suddenly over.

The cat, though, was out of the bag. Before all of France, traditional Christianity had been openly questioned, derided, attacked, and then replaced with reason. Rationality had been set up against dogmatism in the most spectacularly visible of manners—and it was a sight that the Western world, ever since, has never been able to unsee.

## Positive Thinking

Auguste Comte (1798–1857), yet another Frenchman, managed to avoid getting caught up in the French Revolution by rather cleverly being born at the very end of it. His parents—who, somehow, still had their heads on their shoulders—were strict Catholic Royalists, but Comte did not follow in their footsteps. He rejected Christianity, snubbed the reinstated monarchy, and plunged himself instead into the world of secular philosophy.

The switch paid off. By the age of just 28, Comte was teaching his own take on life to some of the best minds around, including Joseph Fourier (1768–1830) and Alexander Von Humboldt (1769–1859)—a fact which is particularly notable given that both men were brilliant scientists already, and both were significantly his senior.

All, however, was not well. Comte had married Caroline Massin (1802–1877) and they were both often desperately unhappy. She was a strong and independent woman—which had played a part in his initial attraction to her—and this, when combined with Comte's instability, jealousy, and paranoia, led to a stormy relationship between the two of them.

Only a year after he began lecturing on his radical new ideas, Comte had a nervous breakdown, with devastating effects for him and his family. Mary Pickering, who has written extensively on Comte, describes just how extreme matters quickly became:

> Sombre and uncommunicative, he would often crouch behind doors and act more like an animal than a human. He still had many fantasies. Every lunch and dinner, he would announce he was a Scottish Highlander from one of Walter Scott's novels, stick his knife into the table, demand a juicy piece of pork, and recite verses of Homer. He often tried to scare Massin by throwing his knife at her. . . . One day, when his mother joined them for a meal, an argument broke out at the table, and Comte took a knife and slit his throat. The scars were visible for the rest of his life.[16]

Remarkably, Massin nursed her undeserving husband back to full health by applying a psychology well ahead of its time. She removed the precautionary protective bars from his windows so that he would feel less patronized; she sat and took a dose of his medicine with him whenever he had to, so that he would feel less like a patient. There were still some significant ups and downs—Comte tried to kill himself by jumping off a Paris bridge and was stopped, somewhat ironically, by a royal guard—but Massin's techniques worked. Before long, Comte was teaching again.

Committing his theory of life, the universe, and everything—a view he called *positivism*—to paper, Comte produced a mammoth six-volume text entitled *Cours de Philosophie Positive*. Although it was methodical and meticulous in its presentation, the *Cours* was also horribly dense and tended to put off all but the most hardened of readers. Fortunately for Comte, he was once again rescued from a potential plunge into obscurity by an inventive young woman. This time it was Harriet Martineau (1802–1876)—from Ambleside, in England.

Martineau had ploughed her way through Comte's *Cours* and was won over by the sheer scope of his vision. She promptly translated it into English, condensing it in the process. The outcome was so much more palatable than the original that it soon became the version of choice for interested readers. Comte himself was so impressed that he had Martineau's text translated back into French again—and advised his compatriots to study her edition instead of his own.

Central to Comte's thesis was his identification of what became known as the Three Stages of human thinking. Martineau's introduction to these stages is as follows:

The law is this: that each of our leading conceptions—each branch of our knowledge—passes successively through three different theoretical conditions: the Theological, or fictitious; the Metaphysical, or abstract; and the Scientific, or positive.[17]

For Comte, there was a clear hierarchy of thought, with theology at the bottom and "positive" science at the top. Note that "Theological" views are assumed to be "fictitious" and that, in the hunt for ultimate reality, humanity must work its way up from such fairy-tale lowlands to the "Scientific" summit. Martineau continues:

In the theological state, the human mind, seeking the essential nature of beings, the first and final causes (the origin and purpose) of all effects—in short, absolute knowledge—supposes all phenomena to be produced by the immediate action of supernatural beings.[18]

Believing in God (or gods) like this is naïve, Comte argues. Real progress is only made once we begin to let go of such childlike ideas:

> In the metaphysical state, which is only a modification of the first, the mind supposes, instead of supernatural beings, abstract forces . . . inherent in all beings, and capable of producing all phenomena.[19]

This metaphysics, warns Comte, is only marginally better than theology. It still believes that there are driving forces and purposes behind life, even if these have now become vague and abstract, rather than godlike. Espousers of pantheism have such a view, for instance—that the Earth, or even the universe, is "alive" as a whole somehow, and has both design and meaning. Comte considers such views to be wholly incorrect but credits them, nonetheless, as being a necessary stepping-stone toward the goal—which is coolly objective science:

> In the final, the positive, state, the mind has given over the vain search after absolute notions, the origin and destination of the universe, and the causes of phenomena, and applies itself to the study of their laws. . . . What is now understood when we speak of an explanation of facts is simply the establishment of a connection between single phenomena and some general facts, the number of which continually diminishes with the progress of science.[20]

In other words, we will always arrive, ultimately, at a mathematical formulation—a law—which allows us to calculate an answer, and then do no more. We should assign this law (and its "answer") no meaning, no purpose, no design, and certainly no designer—it just *is*. And, once we have found it, that particular "branch of our knowledge" is truly "positive"—it is complete.

Religion, under Comte, was defunct. It simply wasn't needed. All conceivable subject matter—even so-called religious experiences—would eventually yield to positivism. Cold and blind equations lay underneath all of existence, and that was that—we just had to find them. Comte was sure that his godless, meaningless, and uncompromisingly rational system was all-sufficient.

Then, in 1846, he changed his mind.

## The Religion of Humanity

Devising a philosophy is one thing; living by it is quite another. Comte discovered this painful lesson twice over—at least.

Firstly, his miserable marriage to Caroline Massin came to an end when, in 1842, she left him for the final time. Although his *Cours* theoretically championed progressive intellectual independence among both men and women, Comte ended up, in reality, longing for a quiet and subservient wife. Massin later wrote "My great crime was to see in you a husband, not a master."[21]

A second incongruity between Comte's theory and his practice was dramatically exposed when he met Clotilde de Vaux (1815–1846). She was a devout Catholic; she was traditionally feminine; she was gentle, and emotional, and delicate—and Comte promptly fell head over heels in love with her. Clotilde then died, within a year of his first letter to her, at the tender age of 31.

Comte was devastated—and became increasingly haunted by the harshness of the universe he had created. He had dismissed all theological and transcendental affectations as small-minded, and supplanted them with cold, hard, mathematical and logical facts—brute facts that came from nowhere and didn't care about anyone or anything. Yet where, in this mechanical scheme, could he find the emotional comfort he now craved? Where was there hope? Where was love?

Echoing his revolutionary forebears, Comte decided that getting rid of religion entirely was not the answer after all—it was more that it needed updating. Inspired by this insight, he devised the Religion of Humanity. Writing at length again, he proposed what he called the positive church—complete with its own secular saints, a priesthood, a catechism, and even a calendar.

Although this new religion sold itself as an intellectual exercise, the whole initiative was, really, a desperate attempt from a desperate man to fix his own broken heart. In a remarkable 1918 journal article, "A Psycho-Analytic Study of Auguste Comte," Phyllis Blanchard made this all too clear:

> "Dear angel," he says, after her death, "I can only adore you by trying to serve better the Great Being in whom I know you are irrevocably incorporated." But this worship for Mme. de Vaux passed all bounds of rationality and became fetishistic in its import. The arm chair in which she was wont to sit became his altar where three times daily he prayed.[22]

This "Great Being" was not God; neither was it some unknown vagueness, as per Robespierre. Instead, it was humanity itself. We were to worship and celebrate ourselves, Comte said, as was made manifest in his prayers to the departed Clotilde—whom he now idolized as everything good about (wo)mankind.

Comte, then, had had two bites of the cherry: the first a robotic, meaningless, laws-only philosophy, and the second a slightly richer version that reintroduced some sort of humanistic bigger picture. Either way, traditional religion was out—and rationality was in.

All in all, Comte taught that humankind is the master of its own destiny; that we must move on from the immaturity of Christian dogma; that science gives us ultimate truth, but allows for no purpose; that we can survive this devastating blow by reflecting upon how magnificent we are (especially Clotilde). In doing so, he left quite the legacy—as is pointed out by literary scholar Tony Davies in his book, *Humanism*:

> The Church of Humanity . . . soon declined, via the usual schisms and internal wranglings. . . . But the informal influence of the cult, with its injunction to "live for others" ("vivre pour autrui," from which we get the word "altruism"), its practice of meditative reflection on the image and example of an idealized Madonna-figure, and its slightly dispiriting vision of a small sphere of human action encompassed on all sides by the vast indifferent presences of nature and history, percolated deeply into the fibre of late-Victorian middle-class thinking.[23]

Comte, on the coattails of the French Revolution, had laid down yet another significant God-versus-science marker for others to mull over—so maybe Draper and White are not the only ones to blame for the conflict thesis after all.

Still, we are not quite done with our investigation—for Davies has provided us with our next major clue. Let's follow that lead, then, and drop in on some Victorian middle-class thinking.

To England!

## Love–Hate Relationships

It is not hard to guess what John Peder Zane's 2007 bestseller *The Top Ten: Writers Pick Their Favourite Books* is all about.

Somehow managing to corral one-hundred-and-twenty-five of the world's foremost authors into sending him their lists, Zane collated their responses and collapsed them down until he was left, in their combined opinion, with the ten best books ever written.[24] All ten have men's names on the cover. One of those men, though, was a woman.

George Eliot (1819–1880)—the writer of tenth-placed *Middlemarch*—was born Mary Ann Evans in Nuneaton, Warwickshire. A brief account of her subsequent life reads very much like a barnstorming novel in its own right.

Growing up in the countryside as the daughter of a devout estate manager, Eliot was quickly deemed unattractive and unlikely to marry. Her only asset of note was her extraordinary mind and, as a bored teenager, she fell under the influence of a nearby group of edgy, progressive, freethinkers— folk who gradually persuaded her to reject her Christian faith, and "discover" herself in the process. Suitably enlightened, she moved to the bright lights and busy streets of London where, having finally come of age, she won a place as assistant editor of the radical and controversial journal, the *Westminster Review*.

The melodrama doesn't stop there. Eliot the Londoner launched herself into a highly public sexual relationship with a brash literary philosopher, George Henry Lewes (1817–1878), with the full permission of said philosopher's wife—for theirs was, scandalously, an open marriage. Despite this racy affair, Eliot saved her deepest and most intimate feelings for a second philosopher, Herbert Spencer (1820–1903), known to all of them—a man for whom true romantic attachment seemed to be a profound difficulty, if not an outright impossibility. It is all quite the soap opera.

And, to top the whole lot off, Eliot also dallied in a lifelong love-hate relationship with the writings and the ideas of a third professional thinker: Auguste Comte.

Bernard J. Paris, professor emeritus of English at Johns Hopkins University, explains more:

> The real crisis in George Eliot's history came not when she broke with Christianity, but when she broke with pantheism, for only then did she have to ask herself if life has any meaning without God. . . . The great question for Eliot, as well as for many of her contemporaries and ours, was, how can man lead a meaningful, morally satisfying life in an absurd universe?[25]

Although she found Comte's first attempt at an answer—the *Cours*— intellectually fascinating, Eliot couldn't help but feel that it offered only an empty and heartless solution. She wanted something more:

What Eliot needed, of course, was a new religion, a religion which would mediate between man and the alien cosmos, as the old religions had done, but which would do so without escaping into illusion[26]

The editor, socialite, and highly successful novelist, therefore, moved on from early Comte to late Comte—and to his Religion of Humanity. Even then, she did not accept it wholesale, but picked out the bits and pieces that she liked. Eliot, then, is a fantastic example of her community—a community that was looking for great big answers, dissatisfied with traditional truths, fascinated by science and natural law, and embarrassed by revealed religion. The conflict thesis feels like it is hovering, here, looking for a suitable landing place.

Would it find one?

## The Specter of Comte

Herbert Spencer was a genuine philosophical heavyweight. Like Comte, his great aim was to draw together a single, overarching scheme of thought that would eventually explain the totality of existence—he called his rival version "synthetic philosophy." Like Comte, he believed that the physical sciences and mathematics were the only route to ultimate truth. Like Comte, he was a religious dissenter who considered conservative biblical Christianity to be outdated nonsense.

An all-rounder who studied and wrote on psychology, politics, physics, economics, and more, Spencer became almost the heartbeat of progressive Victorian culture. He coined the phrase "survival of the fittest," sold possibly as many as a million copies of his works in his own lifetime,[27] and is described by anthropologist Thomas Eriksen as "the single most famous European intellectual" of his era.[28]

Spencer, as mentioned, had a lifelong friendship with George Eliot. They leaned on each other, learned from each other, and admired each other's great intellects. One might even describe them as besotted with one another. The precise nature of their relationship is still much discussed, but we do know that it was never physical—even if Eliot might possibly have wanted it to be.

On one occasion, in fact, the lovelorn Eliot wrote Spencer a letter packed with so much pathos that it could have come straight out of one of her love stories:

I want to know if you can assure me that you will not forsake me, that you
will always be with me as much as you can and share your thoughts and
feelings with me. If you become attached to someone else, then I must
die. . . . If I had your assurance, I could trust that and live upon it. . . . Those
who have known me best have always said, that if ever I loved any one thoroughly my whole life must turn upon that feeling, and I find they said truly.
You curse the destiny which has made the feeling concentrate itself on
you—but if you will only have patience with me you shall not curse it long.
You will find that I can be satisfied with very little, if I am delivered from the
dread of losing it—I suppose no woman ever before wrote such a letter as
this.[29]

Given the extraordinary intensity of both their relationship and their
minds, it is impossible to believe that the two of them did not discuss Comte's
positivism extensively—Eliot was obsessed with it, and Spencer's synthetic
philosophy covered essentially the same topics, making, at times, some almost indistinguishable arguments. The Englishman, however, went to his
grave protesting that Comte had never influenced either his thought or his
work at all; and that he himself was not even slightly a positivist; and that
Comte's analysis was completely wrong, anyway.

That Spencer could make these dismissive claims with a straight—
indignant, even—face is curious indeed. Here, for example, is a quote from
his 1851 book on sociology, *Social Statics*:

Progress, therefore, is not an accident, but a necessity. Instead of civilization being artificial, it is a part of nature; all of a piece with the development
of the embryo or the unfolding of a flower. The modifications mankind
have undergone, and are still undergoing, result from a law underlying the
whole organic creation; and provided the human race continues, and the
constitutions of things remains the same, those modifications must end in
completeness.[30]

Comte could have written that.

And yet, like a schoolboy caught doing something he shouldn't have been,
Spencer continued to insist that he had never even read Comte until three
years after this passage was written and that, when he finally began to, he
thought it was bunk and gave up pretty quickly: "Being an impatient reader,
especially when reading views from which I dissent, I did not go far."[31]

Still, inspired by Comte or not, Spencer is key to our investigation into the origins of the conflict thesis. Was it really all the invention of Draper and White, as seems to be often suggested—or were they merely two more faces in a crowd?

Well, here is Spencer, offering his opinion on the Bible's opening book, Genesis, and its relation to science:

> Ask one of our leading geologists or physiologists whether he believes in the Mosaic account of the creation, and he will take the question as next to an insult. Either he rejects the narrative entirely, or understands it in some vague non-natural sense. Yet one part of it he unconsciously adopts; and that, too, literally. For whence has he got this notion of "special creations," which he thinks so reasonable, and fights for so vigorously? Evidently he can trace it back to no other source than this myth which he repudiates. He has not a single fact in nature to cite in proof of it; nor is he prepared with any chain of reasoning by which it may be established. Catechize him, and he will be forced to confess that the notion was put into his mind in childhood as part of a story which he now thinks absurd.[32]

This is a nuanced passage. In it, the most-read thinker of his generation—which was, of course, also Draper's and White's generation—is telling his audience to grow up. They could have the Sunday school stories of their youth, or they could have the truth. Their thoughts could be led by a naïve, literalistic reading of the Bible, or they could be led by the more mature argumentation of logic and science. Spencer, like the revolutionaries and Comte, was warning of an inherent and unavoidable tension between dogma and reason—a tension that, ultimately, demanded a choice to be made.

And he was not the only member of the English intelligentsia to bang on that particular drum.

## Belligerent Bulldogs and Strangled Snakes

At a mere twenty-five years old, Thomas Henry Huxley (1825–1895) was elected a fellow of the Royal Society. At thirty, he was made professor by the also-very-prestigious Royal Institution. These achievements are made all the more remarkable when we note that Huxley had left school as a youngster due to financial pressures and, from the age of ten onward, had been entirely self-taught.

Skilled in anatomy, physiology, technical drawing, rhetoric, philosophy, and anthropology—with a stint in the navy to boot—Huxley was witty, charismatic, and driven. He was never afraid to put his opinion forward, even when it touched on highly controversial subjects—including, of course, the relationship between progressive science and conservative doctrine. Here he is, in typically bombastic form:

> Extinguished theologians lie about the cradle of every science as the strangled snakes beside that of Hercules; and history records that whenever science and orthodoxy have been fairly opposed, the latter has been forced to retire from the lists, bleeding and crushed if not annihilated; scotched, if not slain.[33]

Huxley was a Victorian celebrity. His fierce support of modern science earned him a nickname—"Darwin's Bulldog"—and he soon became famed for his spiky public debates with pious Christians. An enemy of both Catholicism and Protestantism, he felt that that their outdated creeds—and, indeed, even the newer, more liberal versions of them—simply could not stand up in an era of logic, and experiment, and methodical study:

> I invented the word "agnostic" to describe people who, like myself, confess themselves to be hopelessly ignorant concerning a variety of matters about which metaphysicians and theologians both modern and heterodox dogmatise with the utmost confidence.[34]

Huxley, then, added his voice to those of Robespierre and Comte and Spencer in calling for supernatural doctrines to be dropped—they were infantile, overly prescriptive, and unjustifiable in a scientific age. Yet, despite his outspokenness, even this belligerent and iconoclastic bulldog didn't manage to cause quite as much science-and-religion chaos as another well-known Victorian scientist did, in Belfast, in 1874.

Come on down to the stage, John Tyndall.

## Belfast and the Furious

The British Association for the Advancement of Science (BAAS) was founded in 1831. A powerhouse of academic thought, it boasted the involvement of pretty much anyone and everyone who was making scientific waves at the time. Among those to hold its presidency were William Whewell (1794–1866), who was the first to use the term "scientist"; Richard Owen (1804–1892), who was the first to classify dinosaurs; and Prince Albert (1819–1861), who was the first to marry Queen Victoria.

In 1874, the BAAS appointed John Tyndall (1820–1893), a much-lauded Irish physicist, as its next president. Tyndall, like Huxley, had emerged from relative obscurity to achieve great things in the sciences, including work on the Earth's greenhouse effect, the behavior of glaciers, and the movement of sound through air. Also like Huxley, Tyndall was a brilliant communicator, one who knew how to hold an audience in his thrall. On no occasion would this matter more than in Belfast, on August 20th of that fateful year—when he was due to give the British Association's traditional presidential address.

Tyndall had been building himself up for the event for quite a while, going so far as to write draft material as much as a year beforehand. Traditionally, the address was a rather dry affair; a safe-as-houses, painting-by-numbers lecture which would note some scientific progress made that year, avoid any emotion or controversy, and then give way to opportunities for small talk and networking.[35]

Tyndall, however, had other ideas.

Two years earlier, the Irishman had already annoyed the religious establishment by suggesting that hospital wards be used to test the efficacy of prayer by scientific experiment—thus proving himself to be a potentially provocative participant in the God-and-science conversation.[36] Now, in Ireland, he saw the chance to nail his colors well and truly to the mast—and to do so on one of the grandest stages available.

Despite some of his closer friends warning him to consider his words very carefully—more on this shortly—Tyndall steeled himself to break with tradition and launch into a full-blown, big-picture analysis of all of nature.

The result, it would be fair to say, was dynamite.

Tyndall began his address by lamenting our earliest ancestors' naivety in attributing just about everything to the will of supernatural beings—and then moved on to praise "men of exceptional power, differentiating themselves from the crowd" who dared to "connect natural phenomena with their physical principles."[37] Thanks to these men, he said, Greek and Roman science was born: all was bright and beautiful, and a great deal of pleasing progress was made. At this point, Tyndall paused to ask a question:

What, then, stopped its victorious advance? Why was the scientific intellect compelled, like an exhausted soil, to lie fallow for nearly two millenniums before it could regather the elements necessary to its fertility and strength?[38]

His answer, broadly speaking, was religious dogmatism—with Christianity, in particular, proving to be a major stumbling block:

> Not unto Aristotle, not unto subtle hypotheses, not unto Church, Bible, or blind tradition, must we turn for a knowledge of the universe, but to the direct investigation of Nature by observation and experiment.[39]

Building to a crescendo, Tyndall hammered his main point home one final time:

> All religious theories, schemes and systems which embrace notions of cosmogony, or which otherwise reach into the domain of science, must, in so far as they do this, submit to the control of science, and relinquish all thought of controlling it.[40]

The response, over the next few days and weeks, was nothing short of uproar. In his preface to the sixty-page published version, Tyndall boasted that he had "provoked an unexpected amount of criticism . . . numberless strictures and accusations, some of them exceeding fierce."[41]

Bernard Lightman—one of our aforementioned science-and-religion investigators—has studied Tyndall for decades. He recounts what happened to the physicist in the aftermath:

> Tyndall was denounced in the periodical press . . . a London merchant by the name of C.W. Stokes had sent an inquiry to the Home Secretary, asking whether Tyndall should be imprisoned for blasphemy. . . . The entire British religious world seemed to be against him.[42]

Indeed, Frank Turner—historian, and former second-in-command of Yale University—goes even further: "Probably no single incident in the conflict of religion and science raised so much furore."[43]

So much for painting by numbers.

## Blood on Their Hands

Perhaps we should stop here for a moment, and do a little summing up.

Firstly, we saw that Draper and White appear to have been singled (or doubled) out as the likely perpetrators of our conflict thesis felony—and by the top detectives, no less. It is Draper and White who appear, time and again, in the court documents; it is Draper and White who have spent most of the time in the dock.

Yet the likes of Hébert, Robespierre, Comte, Martineau, Spencer, Huxley, and Tyndall all seem to have blood on their hands, too. Each one of these people has, in one way or another, pitted science against religion, and done so publicly. Each has willingly promoted some form of the conflict thesis to the masses.

And they, as it happens, are not alone either. We have not even mentioned, for instance, the English deists or the German higher critics—both of whom will turn out, in Chapter 7, to be key to the ever-thickening God-versus-science plot. It seems that the list of suspects just goes on and on. So, is ours not a case of two lone gunmen, after all?

Is it actually an act of organized crime?

## The X Club

Thomas Archer Hirst (1830–1892) was, by way of profession at least, a mathematician. What he is most famous for, however, is not his extensive work on geometry, nor his fellowship of the Royal Society, nor even his assorted professorships—it is, instead, his diary.

Hirst's entry for November 6, 1864, reads:

On Thursday evening Nov. 3, an event, probably of some importance, occurred at the St George's Hotel, Albemarle Street. A new club was formed of eight members: viz: Tyndall, Hooker, Huxley, Busk, Frankland, Spencer, Lubbock and myself. Besides personal friendship, the bond that united us was devotion to science, pure and free, untrammelled by religious dogmas. . . . There is no knowing into what this club, which counts amongst its members some of the best workers of the day, may grow, and therefore I record its foundation.[44]

Yes, the "Tyndall" is John Tyndall. Yes, the "Huxley" and "Spencer" are our Huxley and our Spencer. What's more, this night did not go down in history as a one-off meeting of minds—the group of men continued to dine and converse together, on a monthly basis, for the next thirty years. They even gave themselves a name—the X Club.

Ruth Barton, historian and mathematician, knows the X Club better than most—she has written on its formation, members, and influence in impressive detail on multiple occasions. Remarking on the diary entry above, she says:

> Hirst was correct in his guess that the club would become important. Its members were closely associated with the defense of evolutionary theory and the advocacy of scientific, naturalistic understandings of the world; they were representatives of expert professional science to the end of the century, becoming leading advisors to government and leading publicists for the benefits of science; they became influential in scientific politics, forming interlocking directorships on the councils of many scientific societies. James Moore [History Professor at Cambridge and Harvard] describes the club as "the most powerful coterie in late-Victorian science."[45]

The X Club was not a cynical or pragmatic grab for power; neither was it a mere think tank. It was, primarily, a meeting of friends—the members of the group had known each other for years before it first formally convened. In 1851, for example, when Tyndall was nominated for a fellowship of the Royal Society, Huxley signed it off. In 1860, when Huxley's young son died, it was Tyndall who took him to the mountains of Wales to take his mind off things.

It was Spencer who had introduced Huxley to Eliot, and to literature's high society; it was Huxley who had introduced Spencer to the scientific community, and its varied associations. When Spencer got all defensive about his alleged dependence on Comte, he composed an article denouncing positivism—and Huxley and Tyndall, as a personal favor to their friend, promptly added their signatures to it.

As we might expect, then, the three men discussed their views on science and religion at length. Huxley had been president of the British Association four years before Tyndall and had, after much soul-searching, given an uncontroversial lecture. In 1874, aware that Tyndall might choose a more radical approach, he cautioned his colleague not to go overboard. Bernard Lightman details the resulting correspondence between X Club members:

Huxley wrote, "I wonder if that Address is begun, and if you are going to be as wise and prudent as I was at Liverpool. When I think of the temptation I resisted on that occasion . . . I marvel at my own forbearance! Let my example be a burning and shining light to you." Tyndall wrote [to Thomas] Hirst on July 5 that he was amused "to find Huxley expressing anxiety."[46]

The existence, longevity, and influence of the X Club serve to confirm what we might already have suspected—that the likes of Spencer, Huxley, and Tyndall were not just acquainted, but were constantly feeding off one another's work. As Hirst said in his diary, what united them was "devotion to science . . . untrammelled by religious dogma"—and so they continued to promote this view, both internally and externally.

But what about Draper and White?

## Every Thoughtful Man Must Take Part

By 1872, John Tyndall's scientific discoveries and thrilling demonstrations had made him famous enough to merit an invitation to America. It came from the inventor of the electromagnet, Joseph Henry (1797–1878)—who was, at the time, the first ever secretary of the Smithsonian Institute.

Tyndall happily accepted, and headed across the Atlantic to give more than thirty talks across six US cities. He was a major hit: his delighted audiences often numbered over a thousand. Roland Jackson, biographer of Tyndall and one-time chief executive of the British Association, relates just how quickly the Americans took to this bold and dynamic orator:

> His reputation was such that he found himself better known in Philadelphia than in many English cities. Demand for reprints of his lectures—a series of six on the theme of Light—was huge. Printed by the New York Daily Tribune, some 300,000 were sold across America.[47]

Reports of Tyndall's success were wired back over to Britain; Huxley and his pals were "uncommonly tickled" by them. Of most interest to us, however, is something that happened when the tour was virtually over, and Tyndall was preparing to head home. Bernard Lightman again:

> Near the end of Tyndall's visit, his friends organized a farewell banquet in his honor. Held at Delmonico's in New York City on February 4, 1873, the

attendees included Draper [and] White. . . . Andrew Dickson White praised Tyndall as the embodiment of the spirit of scientific research, a spirit sorely needed for the "political progress of our country."[48]

And there, ladies and gentleman of the jury, we have it.

Yes, Draper and White knew John Tyndall. What's more, they knew Thomas Huxley, and they knew Herbert Spencer, too. They wrote to one another, they read one another's material, they appeared in the same publications, and they mixed in the same circles. As such, their ideas and careers served to buttress one another. Tyndall, for instance, cited Draper during his Belfast thunderbolt—and, hearing of this, the chemist encouraged him right back in a letter:

> Stand fast. Your address is doing great good. If you need any help, let me know . . . the friends of science will stand by you in England as they are standing by me in America. Let us all fight shoulder to shoulder in our fighting.[49]

And, while Spencer, Huxley, and Tyndall were undoubtedly major influences on Draper and White, Comte and Robespierre were not entirely absent either. Here is our history of science detective Lawrence Principe putting the clues together once again:

> What makes a sudden appearance [in Draper's writings] is the explicit influence of the positivism of August Comte. Claiming that all systems of metaphysics have fallen into disarray, Draper asserts the need for a new guide for human thought and declares that this "guide is Positive Science."[50]

White, too, was more than a little clued up on the French Revolution and its intellectual fallout—not only had he lectured on it extensively, but he had also amassed his own collection of thousands of books and manuscripts on the topic.[51] Even when we do not include the huge influence of the deists and higher critics on our two men—which shall be explained in full in good time—we can conclude, without much equivocation on the matter, that they did not act alone. In short, they were much more like Newton than like Ward.

Draper and White, then, were not only standing on the shoulders of giants—they were also fighting shoulder to shoulder alongside them. John Tyndall sums up the group mentality well in his review of *Conflict*: "This, in our day, is the 'conflict' so impressively described by Draper, in which every thoughtful man must take a part."[52]

## The Proof Is in the Reading

Still, there remains a rather important question hanging over all this: how is it that Draper and White seem to have been landed with all of the blame? Were they simply the unfortunate fall guys, pushed out by the rest of the mob to take the heat? Why—given that Comte, and Robespierre, and Spencer, and Huxley, and Tyndall, and many, many more were all telling a similar story—do the experts tend to pin the conflict thesis myth-making almost entirely on Draper and White?

Perhaps the answer is a surprisingly simple one. Perhaps, of all the varied contributors and conspirators on show, they were the most effective. Perhaps they were the best of the storytellers. Perhaps their versions—*Conflict* and *Warfare*—of the God-versus-science legend were just in a whole different league to everyone else's.

Well, there is one sure-fire way that we could test this theory.

We could start reading them.

# 3

# Flat Wrong

It's Right in Front of Our Faces • Proceeding by Inquiry • Honest Seekers of Truth • This War of Twelve Centuries • This Preposterous Scheme • Man the Discoverer • A Quintessential Example • At Least Two and At Most Five • Always, Everywhere, and by All • Utter Ignorance of Scripture • Absolutely No Influence • Assailed with Citations from the Bible • I Don't Know How to Get Rid of This Myth

## It's Right in Front of Our Faces

The Cleveland Cavaliers had the first pick in the 2011 National Basketball Association draft—and they used it to add highly rated youngster Kyrie Irving to their ranks. It turned out to be a smart move. Irving won Rookie of the Year in his first season, was an All-Star in his second, and was voted the best player in the All-Star game in his third. By season five, he had guided

Cleveland to a championship—the city's first in any major sport for more than half a century.

Basketball, though, is not the only thing that Irving has made the headlines for.

Recording an episode of the popular podcast *Road Trippin'* with teammates Channing Frye and Richard Jefferson in early 2017, the celebrated athlete suddenly asked them "Do you believe the world is round?"[1]

Frye and Jefferson chuckled, and confirmed that they did. Irving, however, pressed them further:

> I think you need to do research on it. It's right in front of our faces. I'm telling you it's right in front of our faces. They lie to us. . . . what I've been taught is that the Earth is round, but if you really think about it from a land-scape of the way we travel, the way we move . . . can you really think of us rotating around the sun?[2]

Unsurprisingly, their conversation went viral—and the online reactions ranged from despair to puzzlement to delight. Kyrie Irving, sporting super-star, was genuinely arguing that the Earth is flat.

He is not the first to have done so.

## Proceeding by Inquiry

Back in the 1830s, when Draper was starting his career and White was starting his life, Samuel Birley Rowbotham (1816–1884) was busy doing some experiments. By carefully watching flags on barges as they traveled down the Old Bedford Canal in Cambridgeshire, England, he was going to prove that the Earth was a flat disc. Satisfied with his results, he published them (under the pseudonym "Parallax") in *Zetetic Astronomy: Earth Not a Globe*.

John Hampden, for one, was convinced—in fact, he was so sure of Parallax's thesis that he put the equivalent of $75,000 of his family's fortune on the line as a wager. Anyone who thought they could prove the Earth was curved, he said, was welcome to demonstrate it—with proper science, of course—to him and his associates. Such a person, he warned, had better have the money ready in advance—for they would fail.

His challenge was accepted by none other than Alfred Russel Wallace (1823–1913)—an impressively qualified fellow of the Royal Society, who is most famous nowadays for coming up with a well-developed evolution-by-natural-selection model several years before Darwin did.

Hampden and Wallace agreed to carry out the Bedford experiments again, in front of a neutral committee. This time, though, Wallace allowed for atmospheric refraction—a factor which Rowbotham hadn't considered. The new, corrected result was both emphatic and decisive: the Earth was round after all. Rowbotham, presumably, would have been pleased with this new finding—for the word "zetetic" actually means "proceeding by inquiry."

Hampden—the big loser in all this—was furious. Firstly, he complained about the outcome, suggesting that the observers were not neutral at all. When that didn't work, he falsely claimed that he had gone bankrupt, and could not pay off the bet. That failed too. Descending deeper and deeper into the darkness, Hampden wrote a rather bone-chilling letter to Wallace's poor wife:

> Madam, if your infernal thief of a husband is brought home some day on a hurdle, with every bone in his head smashed to pulp, you will know the reason. Do you tell him from me he is a lying infernal thief, and as sure as his name is Wallace he never dies in his bed. You must be a miserable wretch to be obliged to live with a convicted felon. Do not think or let him think I have done with him.[3]

There is no happy ending to this story. Hampden was imprisoned; Wallace got no money.

And yet, despite Rowbotham/Parallax being so soundly proved wrong, his theories were never quite killed off. The Universal Zetetic Society was set up by his supporters in 1901 after his death, and eventually morphed into the International Flat Earth Research Society in 1956. This, in turn, was rebranded The Flat Earth Society, which went online in 2004, and is positively thriving—more than 150 years after Wallace seemed to have got rid of the only reason for its existence.

Today, according to pollsters YouGov, only two in three Americans aged between 18 and 24 are convinced the Earth is round.[4] Celebrities such as rapper B.o.B. and controversial TV star Tina Tequila have added their skeptical voices to that of Irving's. The Flat Earth Society's website receives one hundred thousand visits a month. The whole thing is really taking off.

But why? And how? Are we missing something here?

## Honest Seekers of Truth

The Flat Earth Society offers its own set of answers to the most commonly asked questions about their ideas. Here are some highlights from their web page:

> *What is gravity?* Gravity as a theory is false. Objects simply fall.
>
> *What about pictures from space?* There are a plethora of resources available that show us we can't trust the photographic evidence from organizations such as NASA.
>
> *What does the map of the Earth look like then?* As evidenced by the logo of the United Nations the Earth is a round disk of indefinite dimensions.[5]

This is not only an online movement, however—flat-earthers have begun getting together in real life as well. Meetings have been surprisingly well attended, as reporter Mack Lamoureux discovered:

> "I guarantee that you know flat earthers right now, but they're never going to tell you because they're afraid of being ostracized," Sargent [a prominent figure in the movement] says to me later that evening . . . it's hard not to believe him. If a flat Earth conference in Edmonton, Alberta, [Canada] of all places, can pull in over 200 people at 200 bucks a pop well. . . . I think we may be underestimating the size of the movement.[6]

It just seems bizarre: how has a fringe organization debunked in the nineteenth century suddenly re-emerged to gain such a following? Well, one possible reason is hinted at by YouGov themselves: when they published their survey as an article, they gave it the title "Most Flat Earthers Consider Themselves Very Religious."

Hmm.

In fact, their data was damning on that particular front, for more than half of those who believed in the flat Earth called themselves "very religious."

Those who considered the Earth to be a globe were far less likely—at only 20%—to fall into that category.[7]

Meanwhile, if we hop back to that official Flat Earth FAQ:

> *Is the Flat Earth connected to any religion?* . . . we have no official connection with any established religions. However, it would be impossible to deny the strong historical ties with Christianity by past Presidents of the Society.[8]

This is starting to look rather embarrassing for those of a Christian persuasion. Christine Garwood, who has written extensively on the history of this topic, does not offer much respite—she says that both John Hampden and Samuel Rowbotham saw their geographical views as deeply rooted in their faith: "Like Parallax, [Hampden] was competing for cultural authority . . . his alternative Bible science was the 'true' science, zetetics were the honest seekers of truth."[9]

It seems, therefore, that there is an undeniable relationship between Christianity and the refusal to accept that we live on a globe. Is this not a perfect example, then, of how religion is a natural enemy of science? Is this not precisely the point that Spencer, and Tyndall, and Comte were making in our previous chapter—that dogma closes our minds, and ignores all the evidence?

Well, if flat-earthery is indeed a perfect example of the conflict thesis, then we should expect to find it mentioned by Draper and White, should we not? So, did our two men and their two books engage in any of these God-versus-science arguments about the shape of the planet we live on?

Let's have a look and see.

## This War of Twelve Centuries

It is time, now, to begin our first real deep dive into the twin texts that fooled the world. As we descend further and further, we will hopefully develop a real sense of the mood and style of each volume, as well as that of each writer. We shall set off, in this first instance, with Andrew Dickson White's gigantic work, *Warfare*.

White tended to stick to a specific pattern as he addressed one lengthy topic after another. Each discussion begins with a story of what humanity used to believe in its hopelessly naïve starting state, before showing how science brings us closer to the truth. Along the way, dogmatic religion (usually in the form of one traditional church or another) hovers about as a constant menace: it niggles at, or stalls, or even manages to seriously wound the scientific forces of progress. In the end, however, White makes it clear which side is destined to win out—and it isn't conservative Christianity.

Here we go, then, starting out on *Warfare*'s epic of the flat Earth:

Among various rude tribes we find survivals of a primitive idea that the earth is a flat table or disk, ceiled, domed, or canopied by the sky, and that the sky rests upon the mountains as pillars. Such a belief is entirely natural; it conforms to the appearance of things.[10]

Having ticked that first box, White moves on toward scientific hope:

But, as civilization was developed, there were evolved, especially among the Greeks, ideas of the earth's sphericity. The Pythagoreans, Plato, and Aristotle especially cherished them.[11]

and then he introduces the bad guys:

A few of the larger-minded fathers of the Church, influenced possibly by Pythagorean traditions, but certainly by Aristotle and Plato, were willing to accept this view, but the majority of them took fright at once. To them it seemed fraught with dangers to Scripture.[12]

White goes on to name a host of prominent Christian thinkers who stood firm against the onslaught of science, holding faithfully on to their flat Earth—including this interesting character:

Lactantius referred to the ideas of those studying astronomy as "bad and senseless," and opposed the doctrine of the earth's sphericity both from Scripture and reason.[13]

and this one:

According to Cosmas, the earth is a parallelogram, flat, and surrounded by four seas. It is four hundred days' journey long and two hundred broad. At the outer edges of these four seas arise massive walls closing in the whole structure. . . . The whole of this theologico-scientific structure was built most carefully and, as was then thought, most scripturally.[14]

Lanctantius and Cosmas, White explains, had combined to give the Church's final word on the matter—and they had done so before the end of the sixth century. Drawing on all the relevant Bible passages, they had comprehensively described God's world, and told pious Christians what they were to believe. These Christians, White says, were both impressed and grateful:

Some of the foremost men in the Church devoted themselves to buttressing [the model] with new texts and throwing about it new outworks of theological reasoning; the great body of the faithful considered it a direct gift from the Almighty.

[Cosmas's plan] was accepted by the universal Church as a vast contribution to thought; for several centuries it was the orthodox doctrine, and various leaders in theology devoted themselves to developing and supplementing it.[15]

Then, having established the foolishness of "orthodox doctrine," White encourages his readers not to worry—for science surely forces its way to the surface, even when theology has tried its best to hold it down:

But the ancient germ of scientific truth in geography—the idea of the earth's sphericity—still lived. Although the great majority of the early fathers of the Church, and especially Lactantius, had sought to crush it beneath the utterances attributed to Isaiah, David, and St. Paul, the better opinion of Eudoxus and Aristotle could not be forgotten . . . the sacred theory struggled long and vigorously but in vain. Eminent authorities in later ages, like Albert the Great, St. Thomas Aquinas, Dante, and Vincent of Beauvais, felt obliged to accept the doctrine of the earth's sphericity, and as we approach the modern period we find its truth acknowledged by the vast majority of thinking men.[16]

Christianity, *Warfare* trumpets, was dragged kicking and screaming into modernity by the sheer power of "scientific truth"—entirely against its sacred will. Reluctantly, its biggest names bowed the knee under the combined weight of evidence and reason—for their Bibles were simply not strong enough to support them anymore.

The final death knell of the old Scriptural view was sounded when Christopher Columbus (1451–1506) defied the warnings of the Church to sail over to the West Indies—and, when Ferdinand Magellan's (1480–1521) later expedition went right around the globe, even the most pious flat-earthers had to finally give up:

> The warfare of Columbus the world knows well . . . how sundry wise men of Spain confronted him with the usual quotations from the Psalms, from St. Paul, and from St. Augustine; how, even after he was triumphant, and after his voyage had greatly strengthened the theory of the earth's sphericity . . . the Church by its highest authority solemnly stumbled and persisted in going astray. . . . But in 1519 science gains a crushing victory. Magellan makes his famous voyage. He proves the earth to be round.[17]

Surely now the battle is over—is it not? No, White says, not quite: "Yet even this does not end the war. Many conscientious men oppose the doctrine for two hundred years longer."[18] But it is too late for these faithful few—science will overpower them:

> Then the French astronomers make their measurements of degrees in equatorial and polar regions, and add to their proofs that of the lengthened pendulum. When this was done, when the deductions of science were seen to be established by the simple test of measurement, beautifully and perfectly, and when a long line of trustworthy explorers, including devoted missionaries, had sent home accounts of the antipodes, then, and then only, this war of twelve centuries ended.[19]

His tale told, White sums up his analysis:

> Such was the main result of this long war; but there were other results not so fortunate. The efforts of Eusebius, Basil, and Lactantius to deaden scientific thought; the efforts of Augustine to combat it; the efforts of Cosmas to crush it by dogmatism; the efforts of Boniface and Zachary to crush it by force, conscientious as they all were, had resulted simply in impressing upon many leading minds the conviction that science and religion are enemies.[20]

The point is abundantly clear: dogmatic Christianity, with its naïve reliance on the Bible and stubborn opposition to science, had totally messed up.

Science, with its clarity of thought and willingness to follow the actual evidence, had triumphed. White was teaching the world a valuable lesson.

Such were the flat-earth thoughts of *Warfare*—did *Conflict* concur?

### This Preposterous Scheme

Draper's text may well have been much shorter and snappier than White's, but he still found plenty of room to talk about the shape of the Earth. In fact, his story arc is remarkably similar to his conflict thesis colleague's—and so he, too, begins with early man:

> An uncritical observation of the aspect of Nature persuades us that the earth is an extended level surface which sustains the dome of the sky . . . he seems justified in concluding that every thing has been created for his use—the sun for the purpose of giving him light by day, the moon and stars by night.[21]

The Greeks, though, were smarter than this, and came up with pretty clever scientific arguments for a round earth, says Draper—and then he moves on to a certain Christian thinker:

> Lactantius, referring to the heretical doctrine of the globular form of the earth, remarks: "Is it possible that men can be so absurd as to believe that the crops and the trees on the other side of the earth hang downward, and that men have their feet higher than their heads? If you ask them how they defend these monstrosities, how things do not fall away from the earth on that side, they reply that the nature of things is such that heavy bodies tend toward the centre. . . . I am really at a loss what to say of those who, when they have once gone wrong, steadily persevere in their folly, and defend one absurd opinion by another."[22]

And, having referenced Lactantius, Draper heads dutifully on to Cosmas:

> Perhaps, however, I may quote from Cosmas Indicopleustes the views that were entertained in the sixth century. He wrote a work entitled "Christian Topography", the chief intent of which was to confute the heretical opinion of the globular form of the earth. . . . He affirms that, according to the true orthodox system of geography, the earth is a quadrangular plane . . . that it

is inclosed by mountains, on which the sky rests; that one on the north side, huger than the others, by intercepting the rays of the sun, produces night.[23]

The whole thing is both ridiculous and mildly depressing to our chemist: "Was it for this preposterous scheme–this product of ignorance and audacity–that the works of the Greek philosophers were to be given up?"[24] Still charting the same course as White, Draper duly arrives at Columbus:

Among his friends he numbered Toscanelli, a Florentine, who had turned his attention to astronomy, and had become a strong advocate of the glob- ular form. In Genoa itself Columbus met with but little encouragement. He then spent many years in trying to interest different princes in his pro- posed attempt. Its irreligious tendency was pointed out by the Spanish ecclesiastics, and condemned by the Council of Salamanca; its ortho- doxy was confuted from the Pentateuch, the Psalms, the Prophecies, the Gospels, the Epistles, and the writings of the Fathers—St. Chrystostom, St. Augustine, St. Jerome, St. Gregory, St. Basil, St. Ambrose.

and then at Magellan, whose ship:

had accomplished the greatest achievement in the history of the human race. She had circumnavigated the earth. The San Vittoria, sailing west- ward, had come back to her starting-point. Henceforth the theological doc- trine of the flatness of the earth was irretrievably overthrown.[25]

And there we have it—the same tale, told by two eminent authorities, in two bestselling books. Bible-thumping theologians, said White, had "arrested the normal development of the physical sciences for over fifteen hundred years." Draper's synopsis put it even more succinctly:

Scriptural view of the world: the earth a flat surface . . .
Scientific view: the earth a globe.

Point made.

## Man the Discoverer

Such was the force of Draper's and White's flat-earth account that it soon be- came the standard storyline, the definitive discussion, the normal narrative.

When the next generation or two of historians or commentators wrote on the topic, they simply repeated or embellished the *Conflict–Warfare* account as they saw fit. This often applied even to the most highly respected and best qualified of writers—including Daniel J. Boorstin (1914–2004).

Boorstin enjoyed twelve fine years as the librarian of the US Congress. He was a prolific author, penning more than twenty historical works, including *The Americans: The Democratic Experience*—for which he won the 1974 Pulitzer Prize. A big-hitter on the intellectual stage, he was awarded honorary degrees by universities from all over the world. In short, he was considered to be a man who knew what he was talking about.

One of his most popular titles, *The Discoverers*, was published in 1984. We get our first clue that the issue of a flat Earth might feature as early as the introduction: "My hero is Man the Discoverer. The world we now view from the literate West . . . had to be opened to us by countless Columbuses."[26] And, when the seemingly inevitable comes, there is a familiar name attached to it. Chapter Fourteen—"A Flat Earth Returns"—begins like this: " 'Can anyone be so foolish,' asked the revered Lactantius, 'the Christian Cicero,' whom Constantine chose to tutor his son, 'as to believe that there are men whose feet are higher than their heads . . . ?"[27]

From here, Boorstin goes on, carefully tracing the footsteps of the two men who wrote a century before him:

> To avoid heretical possibilities, faithful Christians preferred to believe that there could be no Antipodes, or even, if necessary, that the earth was no sphere. Saint Augustine, too, was explicit and dogmatic, and his immense authority, compounded with that of Isidore, the Venerable Bede, Saint Boniface, and others, warned away rash spirits. The ancient Greek and Roman geographers had not been troubled by such matters.[28]

"Antipodes," by the way, were hypothesized people who lived below the equator. Their existence was hotly debated among thinkers of all persuasions, Christian or not—indeed, it resembled the modern debate about extraterrestrials. While this argument was a separate one from the shape of the planet itself, the two were clearly related, and were often talked about alongside one another. Still, getting back to Boorstin's book, we soon happen across another recognizable character:

> It was a fanatical recent convert, Cosmas of Alexandria, who provided a full-fledged *Topographia Christiana*, which lasted these many centuries to the dismay and embarrassment of modern Christians. . . . In his very first

book he destroyed the abominable heresy of the sphericity of the earth. Then he expounded his own system, supported, of course, from Scripture.[29]

Eventually, Columbus came along and sorted the whole mess out. The Greeks, it was discovered, had been right all along—and, crucially, they had been right without the aid of the Bible.

## A Quintessential Example

Boorstin is not the only more modern thinker to follow Draper and White. Their story shows up all over the place as a useful teaching tool—one which calls us to trust science, and not religion, when it comes to uncovering the truth about our world. Here, for instance, is *Astronomy* by John Fix—a much-used college textbook, which is now in its sixth edition:

> Part of the reason for the loss of Greek astronomical knowledge can be attributed to the antagonism of the early Christian church to many of the features of Greek astronomy. Many church leaders thought that ideas such as the sphericity of the Earth contradicted descriptions of the universe found in the Scriptures. Rather than accept the spherical shape of the Earth and the celestial sphere, some Christian scholars such as Lactantius and Kosmas argued that the Sun, after sunset, travelled around the horizon toward the north and then east to rise again in the morning.[30]

Any number of alternative volumes could have been chosen to make this same point; the *Conflict* and *Warfare* version of events is pretty much everywhere. Historian Edward Grant, a distinguished medievalist, confirms as much:

> The flat-earth theory that was attributed to the Middles Ages gradually became embedded in both popular literature and intellectual thought. Since approximately 1870, the flat-earth theory has become very nearly a quintessential example of the backwardness of the Middle Ages . . . one author placed Columbus before a commission at Salamanca that is made to terrorize Columbus with the following chilling lines: "You think the earth is round, and inhabited on the other side? Are you not aware that the holy fathers of the church have condemned this belief?"[31]

In fact, we can go further—for not only does it inhabit "popular literature and intellectual thought," but the legacy of Draper and White can even be found on Twitter. In 2016, Neil deGrasse Tyson—physicist, science communicator, and internet favorite—rightly decided that it was time to take on the new wave of flat-earthers, and clear up their misinformation for the benefit of a whole generation. He tweeted a gently corrective nudge aimed at the aforementioned rap artist B.o.B: "Duude—to be clear: Being five centuries regressed in your reasoning doesn't mean we all can't still like your music."[32] When a reader queried his history, and mentioned folk knowing about the curvature of the Earth far earlier than "five centuries" ago, Tyson clarified his original statement: "Yes. Ancient Greece—inferred from Earth's shadow during Lunar Eclipses. But it was lost to the Dark Ages."[33]

Floating around cyberspace, then, is Grant's "quintessential example" of dogmatism stifling science, and of Christendom bringing about Tyson's "Dark Ages." Christians, rejecting the wisdom of the Greeks, had effectively stuck their fingers in their ears. Then, not content with that, they did their best to stop everyone else from listening, too. As Draper himself put it: "Catholicism had irrevocably committed itself to the dogma of a flat earth."[34]

Well, it is a good cautionary tale. Blind faith in Bronze Age holy writings won't get anyone all that far—but, if they start doing science instead, they might yet become one of Boorstin's discoverers. Before we move on with our lesson duly learned, however, there is probably something worth double-checking about all this.

Namely, this: is it actually true?

## At Least Two and at Most Five

It is entirely conceivable, of course, that our earliest ancestors really did think that they were living on a flat disc, or in a box, or under a solid sky, or all three—because it does, at first simplistic glance, very much look like we are. So far, so good, then, for Draper and White. Similarly, our duo are right about the Greeks and Romans—for they did indeed use observations and mathematics to deduce the shape of the planet we all call home.

Thanks to clever scientific reasoning about the Earth's shadow on the moon, the disappearance of ships over the horizon, the changing lengths and angles of shadows in different cities, and the varied positioning of stars as one moved over its surface, the best thinkers from around 500 BC onward had conclusively nailed down the sphericity of our world. The likes of Pythagoras, Aristotle, and Euclid had got it all mostly figured out a few centuries before Jesus turned up—and there was, even during their time, a broad consensus about the matter.

It is at this point, though, that things start to go horribly awry for *Conflict* and *Warfare*. To see why, it is perhaps best to introduce a new voice into the proceedings: that of Jeffrey Burton Russell, professor emeritus of history at the University of California and author of the ominously titled *Inventing the Flat Earth*. Russell gets straight into the fray with his preface:

> The almost universal supposition that educated medieval people believed the earth to be flat puzzled me and struck me as dissonant when I was in elementary school, but I assumed that teacher knew best and shelved my doubts.[35]

He then continues with the following bombshell: "By the time my children were in elementary school, they were learning the same mistake, and by that time I knew it was a falsehood."[36]

Wait a second: a "falsehood"? Can Russell support such a radical counterclaim—one that undermines not only Draper and White, but other established intellectuals like Boorstin, and Fix, and Tyson—with any actual evidence? Well, he seems to think so:

> A few—at least two and at most five—early Christian fathers denied the sphericity of earth by mistakenly taking passages such as Ps. 104:2–3 as geographical rather than metaphorical statements. On the other side tens of thousands of Christian theologians, poets, artists, and scientists took the spherical view throughout the early, medieval, and modern church. The point is that no educated person believed otherwise.[37]

Oh. "At least two and at most five" is markedly different to White's "great majority of the early fathers of the Church." Did Russell somehow miss the memo that the others all got? Who is right here? Who is wrong?

Russell, somewhat promisingly for his cause, can boast an ally in Allan Chapman, historian of science at the University of Oxford:

> Let us be clear about one thing: no medieval scholar of any worth thought that the Earth was flat . . . . One needs only to read the astronomical

literature of the Middle Ages to realise that the spherical nature of the earth, about 6000 or 8000 miles across, was standard knowledge, and taught to university students from Salamanca to Prague . . . . Inherited in unbroken succession from the Greeks, in fact, taught by the Venerable Bede to the young monks of Jarrow Abbey in AD 710, and encapsulated in John of the Holy Wood (Johannes de Sacrobosco's) Latin textbook *De Sphaera Mundi* ("On the Sphere of the Earth") of c. 1240.[38]

Lesley B. Cormack—yet another historian of science, this time in Alberta, Canada—is also on Team Russell:

If we examine the work of even early-medieval writers, we find that with few exceptions they held a spherical-earth theory. Among the early Church Fathers, Augustine (354–430), Jerome (d. 420), and Ambrose (d. 420) all agreed that the earth was a sphere.[39]

Later in the same essay, Cormack says that between AD 600 and 1300: "every important medieval thinker concerned about the natural world stated more or less explicitly that the world was a round globe, many of them incorporating Ptolemy's astronomy and Aristotle's physics into their work."[40]

She goes on to name Thomas Aquinas (d. 1274), Roger Bacon (d. 1294), John Sacrobosco (1195–1256), and the archbishop of Cambrai, Pierre d'Ailly (1350–1410) as famous and influential Christians who wrote, freely and openly, about the globular form of the world.

It doesn't end there: medieval bestsellers such as Dante Alighieri's *Divine Comedy* (1320), John Mandeville's *Travels* (c. 1360), and Geoffrey Chaucer's *Canterbury Tales* (c. 1400) all treat the Earth, quite matter-of-factly, as a ball. This is important because, as Cormack herself explains, it tells us that even the non-scientific audiences of the time were fully aware of the truth.

If it is really the case, however, that the doctrine of a spherical Earth was an unquestioned constant from the time of the Greeks onward, and that the Church actively believed and taught it throughout that Middle Ages and beyond, then why did *Conflict* and *Warfare* both insist, so confidently, that the exact opposite was true?

Moreover, why did the very-well-read Boorstin agree with them? Why do so many textbooks, even today, side with Draper and White rather than Russell? Why does Neil deGrasse Tyson's Twitter feed say that Greek conclusions were "lost to the Dark Ages"—until, presumably, the persecuted-but-unbroken Columbus came along? Why does the Church keep getting the blame for a crime that it didn't commit—indeed, a crime that no one committed?

The answer, it would seem, lies with Russell's "at least two."

## Always, Everywhere, and by All

Lactantius and Cosmas have a lot to answer for: it is, essentially, on them that Draper and White and Boorstin and Fix all hang their story. Sure, other saintly names are mentioned, but no one else is ever actually quoted—only these two. Upon investigation, it turns out that there is a rather good reason for this: there is, to put it simply, nobody else to quote. Cosmas and Lactantius, and they alone, are the sources for the world-famous Christian flat Earth.

Still, if Lactantius and Cosmas were revered as messengers from God—if they were the authorities looked to by theologians, priests, and lay believers everywhere for hundreds of years afterward—then they would be enough to build a case upon. Their teachings could genuinely be viewed as representative of Christianity as a whole, and the Draper and White thesis would hold firm. White suggests as much in *Warfare*:

> In summing up the action of the Church upon geography, we must say, then, that the dogmas developed in strict adherence to Scripture and the conceptions held in the Church during many centuries "always, everywhere, and by all," were, on the whole, steadily hostile to truth.[41]

Let us give White the benefit of the doubt for a moment: if the doctrines developed by Lactantius and Cosmas "in strict adherence to Scripture" really were believed to be true "always, everywhere, and by all," then his overall argument—and Draper's, and Tyson's, and so on—clearly stands.

But were they believed "always, everywhere, and by all"? Were these two geographical commentators universally influential? Did they set the holy course for everyone else to follow? Had Catholicism at large accepted their teachings and "irrevocably committed itself to the dogma of a flat earth," as per Draper?

It's time to find out.

## Utter Ignorance of Scripture

Lucius Caecilius Firmianus Lactantius was born in Africa sometime around AD 245 to pagans. He ascended to the lofty role of head of rhetoric under the Emperor Diocletian (245–313), before converting to Christianity nearer to AD 300. Diocletian was about to begin a period of brutal persecution across the empire—one which would see churches destroyed, Scriptures burned, and many Christians mercilessly executed. Lactantius saw it all coming—so he resigned his post and fled to Gaul.

Outlasting the vicious attacks on his spiritual brothers and sisters, he found himself back in favor once Constantine (272–337) took over as emperor in AD 306. The atmosphere had changed, and Christianity was no longer under threat—and, as Boorstin rightly observed, Lactantius was duly appointed as tutor to Constantine's son.

A gifted thinker, Lactantius soon became renowned, even during his own lifetime, for his aesthetically pleasing prose and persuasive argument. Here is a taste of his work—a passage made all the more poignant when one remembers that he had seen much Christian suffering in his life:

> Religion is to be defended, not by putting to death, but by dying; not by cruelty, but by patient endurance; not by guilt but by good faith . . . if you wish to defend religion by bloodshed, and by tortures, and by guilt, it will no longer be defended, but will be polluted and profaned.[42]

He was also convinced—just as Draper and White wrote all those years ago—that the Earth really was flat.

One might think, then, that this famous theologian's insistence upon terrestrial flatness would have carried far and wide. One might think that the ever-sheeplike laity would hang on his every word, and write off any spherical ideas they might have heard elsewhere as blasphemous lies. In reality, however, that is not how matters played out.

Lactantius was well-known, and skilled in rhetoric, yes—but he was hardly a respected doctrinal authority. So confused and muddled was some of his theology, in fact, that he mistakenly taught that Jesus and the devil were twins, that Jesus was an angel created by God, and that God had been forced to create evil in the world because His hands were tied by the laws of logic. Unsurprisingly, such beliefs saw him written off as a heretic after his death.

The *Catholic Encyclopedia*, which we might expect to be his strongest supporter, says this:

> The strengths and the weakness of Lactantius are nowhere better shown than in his work. The beauty of the style, the choice and aptness of the terminology, cannot hide the author's lack of grasp of Christian principles and his almost utter ignorance of Scripture.[43]

This is no after-the-events apology written by an embarrassed modern contributor, either—for St. Jerome (c. 347–420) had made a similar point more than 1500 years ago: "Lactantius has a flow of eloquence worthy of Tully: would that he had been as ready to teach our doctrines as he was to pull down those of others!"[44]

It is not just that people listened to Lactantius about the flat Earth and then, after much discussion, had decided that he was wrong—it is more that his geographical views never appeared on their radar in the first place. Geoscientist Malcolm Reeves spells out the rather uncontentious reality of the situation:

> Starting with Bede (seventh–eighth centuries), a consistent exposition and defense of the sphericity of the earth was clear in Western Europe and made its way into university teaching. Nobody in the Middle Ages took notice of Lactantius's rejection of the sphericity of the earth.[45]

Lactantius, having effectively disappeared after his death, was later resurrected during the explosion of interest in classical authors which occurred in Europe in the fifteenth and sixteenth centuries. By then, however, any chance of him having any impact whatsoever on either doctrine or dogma was long gone. He was a novel curiosity, celebrated by a handful of Renaissance scholars for the quality of his writing—but nothing more. In making him a spokesperson for the whole of Christendom about the shape of the Earth, then, Draper and White—and Boorstin, for that matter—have got things horribly wrong.

Surely they couldn't make the same mistake twice, though—could they?

## Absolutely No Influence

Cosmas Indicopleustes (d. 550) was ethnically Greek, born in Egypt, and earned his surname—literally "sailed to India"—for his traveling exploits. Like Lactantius, he was an adult convert to Christianity and, also like Lactantius, he (mis)used his Bible to make a grand old mess of geography.

Against essentially every other thinker before or since, Cosmas decided that a couple of passages in the book of Hebrews about the heavenly tabernacle—a temple-like structure in which God dwelled and was worshiped—were not about spirituality, but were instead a physical description of the cosmos. This entirely bonkers analysis led him to sketch out the box-shaped world that Draper, White, Boorstin, and the textbooks seem to think was adopted "always, everywhere, and by all."

But it wasn't.

Here is Russell again:

> The influence of Cosmas's blundered effort on the Middle Ages was virtually nil. . . . Cosmas was roundly attacked in his own time by John Philoponus (490–570) . . . after Philoponus, Cosmas was ignored until the ninth century, when the Patriarch Photius of Constantinople again dismissed his views. In Latin, no medieval text of Cosmas exists at all. The first translation of Cosmas into Latin, his very first introduction into western Europe, was not until 1706. He had absolutely no influence on medieval western thought.[46]

To press Russell's point home, it is worth looking at exactly what one prominent Christian teacher he mentioned—John Philoponus—actually had to say about Cosmas:

> If certain people, owing to the uneducated state of their soul, cannot attain to what has been said and are troubled about the way the facts are put together, silence will help them to cover up their own ignorance. And let them not tell lies about God's creation out of their own lack of experience and the slowness of their mind. . . . Some people's saying that it [the sun] is carried by the north winds to return to the east, being hidden by very high mountains, was an ancient and foolish notion held by some which deserves the laughter befitting it.[47]

Ouch.

Perhaps we can leave it to historian of science Pablo de Felipe to hammer the final nail into Draper's and White's everyone-listened-to-Cosmas coffin:

> Interestingly, and contrary to the impression commonly left after the rediscovery of Cosmas in the early 18th century, his work was not the beginning or even the pinnacle of flat-earth cosmological influence among Christians. It was rather the opposite; this most elaborate defense of the flat earth seems to have brought the discussion to its end. As far as we can track in the extant Christian texts of late Antiquity and the early Medieval period, there seem to be no followers of Cosmas.[48]

So, in actual fact, the Church never believed in a flat Earth, never taught it, was never in the need of correction on the matter—and the two poster boys of the *Conflict* and *Warfare* grand narrative, Lactantius and Cosmas, were written off by Christendom as weirdos who were barking up entirely the wrong tree. Draper and White, then, weren't just slightly off course—they were on a different planet altogether.

One loose end remains, however. For, if no one at Columbus's time thought the Earth was flat, then why does the Spanish sailor show up in the story at all? Why did White write the "warfare of Columbus the world knows well"? Why does Draper have the Spaniard warned by the ever-dogmatic Church of his proposed journey's "irreligious tendency"?

The answer is almost unbelievable: the idea first seems to have turned up in the American mainstream in a *novel*.

## Assailed with Citations from the Bible

Washington Irving (1783–1859) is most famous for his short fairy tales "Rip Van Winkle" and "The Legend of Sleepy Hollow." He also loved to write historical pieces—or, more accurately, pseudo-historical pieces. His *A History of the Life and Voyages of Christopher Columbus* (1828) had originally been planned as straight history—but the novelist had found his desired sources to be either wholly non-existent or rather less dramatic than he had anticipated. Historian of science Darin Hayton explains what then ensued: "Irving embellished. He wrote what should have happened, what surely did happen even if the evidence had since disappeared. He did what historians have been doing since Herodotus: he made it up."[49]

In his quest to write a good story, Irving played around with the true events—that Columbus had indeed been challenged, but only because his distance calculations were, quite correctly, considered suspect—and decided to turn what he knew was merely a mathematical debate into an entirely ficti-tious science-versus-religion showdown:

Columbus was assailed with citations from the Bible and the Testament, the book of Genesis, the psalms of David, the Prophets, the epistles, and the gospels. To these were added expositions of various Saints and reverend commentators, St. Gregory, St. Basil and St. Ambrose . . .

wait for it . . .

. . . and Lactantius.[50]

None of this stand-off took place, except in Irving's imagination—but his readers took it to heart nonetheless. It is particularly instructive to compare this paragraph from *Columbus* to one written by Draper, in *Conflict*, 50 years later:

> [Columbus's journey's] irreligious tendency was pointed out by the Spanish ecclesiastics, and condemned by the Council of Salamanca; its orthodoxy was confuted from the Pentateuch, the Psalms, the Prophecies, the Gospels, the Epistles, and the writings of the Fathers—St. Chrysostom, St. Augustine, St. Jerome, St. Gregory, St. Basil, St. Ambrose.[51]

Draper, incidentally, gives no footnote for this bold assertion—but it would hardly take a Hercule Poirot or a Miss Marple to deduce its origins.

And, just to sum up the outcome of all this, here is Hayton once more:

> By the time Andrew White wrote [*Warfare*] Columbus's struggles to overcome a medieval Church that believed in a flat earth had become historical fact. Historical truth had surrendered to truthiness. . . . Despite decades of historical work and dozens of articles and textbooks and, more recently, blogposts, the Columbus myth is alive and well.[52]

The flat Earth account, then, is a God-versus-science myth, alive and well today, set on its way by Draper and White.

And it is far from being the only one.

## I Don't Know How to Get Rid of This Myth

Our first deep Draper-and-White dive has been an enlightening one. We have seen the words of *Conflict* and *Warfare* for ourselves. We have seen how both men told a simple story that was easy to follow, and also how the overwhelming majority of the world's commentators have since been persuaded by it.

Our pair spelled out, in quite some style, a moral tale about science and religion: the Greeks had deduced the shape of our planet using science; Christianity had universally rejected sphericity on the basis of Scripture; Lactantius and Cosmas had led the holy charge against reason, with the

Church blindly following along; Columbus, in the face of persecution, had finally rescued us from dogma.

For the last 150 years, coursebook writers, librarians of Congress, and tweeting astrophysicists have continued to remind us all of this important story, and of its even more important core lesson: that dogmatic religion leads us to nonsense, and that rational science leads us to truth. We have to pick a side; and that side, without doubt, should be science.

But Draper's and White's moral tale is not true. It is a mixture of exaggerations, dishonest emphases, misunderstandings, and outright lies— it has been consigned to the rubbish heap by scores of careful historians of science.

Yet still, it thrives—in our schools and universities, as well as on our streets, and in our heads. Russell, in his book *Inventing the Flat Earth*, can be found lamenting the current situation: "Many authors great and small have followed the Draper-White line down to the present. The educated public, seeing so many eminent scientists, philosophers, and scholars in agreement, concluded that they must be right."[53] And this, sadly, is not an isolated case. Here is the impressively titled Katherine Park—the Samuel Zemurray, Jr., and Doris Zemurray Stone Radcliffe Research Professor of the History of Science, Emeritus—giving an interview to the Harvard Gazette in 2011:

> "It was a 19th century myth," said Park, "like that before Christopher Columbus everyone thought the world was flat. People are absolutely wedded to a view that says 'We are modern, and they were stupid' . . . it just kills me. I don't know how to get rid of this myth."[54]

Had Draper and White only struck once—by indoctrinating folk with the flat Earth fallacy—then perhaps the conflict thesis would not have caught on in the way that it has. But they did not strike just once—for the pervasive myth that Park is upset about, and that she doesn't know how to get rid of, also came from the pages of *Conflict* and *Warfare*. So what was it?

Well, that will be the topic of our next chapter.

# 4

# Walnuts for Brains

### Better Biologists?

By the time she was rushed into California's Riverside General Hospital on the 19th of February 1995, Gloria Ramirez was already dying. Her advanced cancer had taken its toll, and she was now suffering from a combination of heart, kidney, and respiratory problems. After their initial efforts to stabilize her failed, the staff switched to emergency measures and prepared to shock her heart back into action. As they sheared off her clothing, they found a thin, oily film all over her skin.

This curious detail would have been quickly forgotten if Ramirez's case had proceeded as normal—but it didn't. Soon, several of those attending her began to pick up a garlicky aroma in the room. Susan Kane, the nurse closest at hand, drew some blood from Ramirez's arm—and noticed that it smelled like ammonia. She passed the syringe to Julie Gorchinsky, who spotted tiny white-brown crystals suspended within the sample. Maureen Welch, a third team member, sniffed at it, checking for the familiarity of anti-cancer medication, but recognized nothing. Everything about the situation seemed somehow wrong.

Suddenly, Kane lost consciousness and crumpled to the ground. Gorchinsky fell too—and then stopped breathing. As the chaos kicked in around her, Welch briefly blacked out. When she came to again, her arms and legs were in violent spasm, and she was unable to control her movement. At this point the doctor in charge, Humberto Ochoa, decided he had better do something, and quick.

The three women were taken off for treatment; the ER was evacuated; patients and carers were moved out into the carpark. A couple of hardy souls were asked to stay with Ramirez in a last-gasp attempt to save her life, but it was to no avail. Less than hour after she arrived, the 31-year-old tragically passed away.

Matters outside, meanwhile, continued to be troubling. By the time it was all over, some twenty-three hospital staff had required their own medical attention. Five were kept in overnight—Gorchinsky remained in intensive care for a further two weeks. Even after her release, she never fully recovered: the unfortunate nurse went on to suffer from liver and pancreas complications, and then the bones in her legs began to die from the inside out—leaving her unable to walk without crutches.

Naturally, there arose an overwhelming demand to know what on earth had actually happened that night. What were the various smells? What was the oil on Ramirez's skin? How had so many people been affected in such a short time in such bizarre ways? Could this happen again? If so, how might staff be better protected?

Since these events, several official bodies have had a crack at explaining what went on—but to this day, a quarter of a century later, we still have no definite conclusions. Hazmat teams could not trace any dangerous chemicals on the site. Coroners found nothing problematic in either Ramirez's body tissues or her blood.

In the last two decades, the closest anyone has come to a workable theory is that the young woman had been self-medicating with a dubious painkiller brewed out of cleaning fluid—and that this chemical, when combined with the oxygen she was being given, had somehow reorganized itself into a form of nerve gas.[1] It should be noted, however, that the required reaction has never been observed by anybody under anything approaching real-world circumstances—and so the case, if we are being honest, remains open.

The strange and worrying tale of Gloria Ramirez is but one of many still unsolved medical mysteries, most of which can be quickly discovered with a simple online search. Alongside them, however, sit plenty of far more mundane physiological puzzles: we don't know, for instance, why we blush. Likewise, we are uncertain about why we yawn; or why we pick our noses; or even why we have fingerprints.

And, at the deeper end of the subject, where more profound truths are thought to dwell, there is a veritable ocean yet to be charted. Perhaps the most perplexing of these harder questions is the so-called problem of consciousness—how on earth does our brain, which is an electrochemical lump of cells, give rise to the universal and undeniable sense that we really are, somehow, a person?

The point of all this—of Ramirez, of blushing, and of personhood—is a humbling one: our understanding of the human body is a long, long way from being complete. There is much that we simply do not understand about ourselves and the way we work—and often what we think we do know is called into question as new occurrences, discoveries, or test results are brought to light.

Yet why is this the case? One might have thought that, of all the sciences out there, the one most involved with looking after ourselves would also be the one we were best at. After all, aren't we supposed to be finely honed survival machines? Shouldn't medicine have always come top of the list for time, for thought, for effort, and for funding? Why, then, are there still so many unknowns?

Let's put it more bluntly: shouldn't we be rather better at biology than we are?

## A Great Start

Most stories from the history of science tend to start off (for better or worse) with the Greeks, but medicine is different—for we can, without too much effort, get much further back than that. We have in our possession, for instance, an Egyptian medical document that was written in 1500 BC—and it may well be a copy of an original text another thousand years older than that.

The product of an unknown author, this intricate and remarkably scientific document is, therefore, emphatically pre-Greek: indeed, it had already been devised, copied, and used for hundreds of years before Plato's great-great-grandparents had considered making their relationship a little less . . . well . . . Platonic. Rather gruesome at times, the pamphlet describes no fewer than forty-eight separate injuries, as well as how to treat them—and it makes fascinating reading.

Case number four in this guide—now known as the *Edwin Smith Papyrus*—has the charming title "Head Injury with Open, Displaced, Elevated, Skull Fracture." The writer notes that, in this particular instance, the doctor is likely to notice the brain "shaking under his fingers" and that the sufferer "bleeds through his nostrils." Here is the overview:

> If you should examine a man for a slash wound in his head, that has penetrated to the bone, splitting his brain-case, you must palpate his wound . . . you should not bandage it. Put (him) to the ground, upon his bed under observation and support until the (critical) period of his injury passes. His treatment: it is sitting (upright). Make for him: two supports of brick until you know he has reached a turning point. Then, you have to put ointment on his head, and soften/massage the area of the back of his neck together with both his shoulders.[2]

Ancient human biology, then, was quite respectably advanced—especially in comparison to the likes of physics or chemistry.

And, when the Greeks did then turn up and do their thing, it got even better.

## Keeping the Momentum Up

No tour of our topic, of course, could possibly bypass Hippocrates (b. 460 BC). Nicknamed the "father of medicine" (although our earlier Egyptian doctor might feel a little hard done by on that front), he helped to pioneer a holistic, evidence-based approach to healing. His system included taking the medical histories of the sufferer and their family, examining their place of work, asking about their general mood, and even working through—by hand—the newly deposited contents of their bowels. What's more, the Hippocratic school also incorporated a clear moral dimension—physicians took an oath not to do any bodily harm and not to exploit their patients in any way. Doctoring was to be a vocation.

After Hippocrates, Aristotle took up the reins. His modus operandi, for most of his vast philosophy, was to observe the world directly and then think about whatever he saw. His eyes were more on the Earth than they were on the heavens, and this could pay dividends when it came to the dirty and messy world of sickness. Dissecting animals, for instance—as Aristotle probably did, and others following him certainly did—offered new insights as to what might be happening underneath the skin of a puzzled and poorly patient.

Then, with that hands-on groundwork laid, came Herophilus of Chalcedon (c. 335–260 BC) and Erasistratus of Ceos (c. 304–250 BC). By taking the next logical step and cutting up human beings, they made a whole range of revolutionary discoveries. Herophilus identified the eye's cornea and retina; he distinguished between veins and arteries; he analyzed the liver; he worked out some of the functions of both male and female reproductive organs. Erasistratus built on all this, noting the one-way passage of blood through heart valves and identifying the heart as a pump.[3] And we are still, by the way, half a millennium before Christendom really kicks in.

The last of the great Greek surgeons was Galen (b. AD 129). Galen's job—which he took up for fun, and not because he needed a living—was to (try to) put gladiators back together again after they had been hacked, stabbed, impaled, or bashed to bits by a rival. This was, unpleasantly, a sure-fire way to get to know first-hand how the human body did (and didn't) work.

Crucially, Galen kept a detailed and almost day-to-day record of his practice: he generated copious amounts of notes, published multiple booklets, drew up scores of schematic sketches, and updated older theories on plenty of bodily processes. All of this know-how was based, one way or another, on his direct personal experience. It was, as such, invaluable.

Long before the Church came into any kind of power, then, the scientific discipline of medicine was in excellent shape. It could boast millennia-old meticulous manuals, the use of patient histories and background checks, a germ of psychology, anatomical theories based on dissection—and an admired, well-paid profession which was sworn to care for, protect, and preserve human life.

Biology, we can assuredly say, was in fairly fine fettle.

But would it last?

## A Change of Tune

When medicine makes it first appearance in *Warfare*, things immediately feel rather familiar:

> In those [earliest] periods when man sees everywhere miracle and nowhere law,—when he attributes all things which he can not understand to a will like his own,—he naturally ascribes his diseases either to the wrath of a good being or to the malice of an evil being.[4]

We are being treated here, of course, to White's usual trope—that our naively religious ancestors were horribly irrational but were, happily for us all, succeeded by the wonderfully scientific Greeks:

> Five hundred years before Christ, in the bloom period of thought—the period of Aeschylus, Phidias, Pericles, Socrates, and Plato—appeared Hippocrates, one of the greatest names in history. Quietly but thoroughly he broke away from the old tradition, developed scientific thought, and laid the foundations of medical science upon experience, observation, and reason so deeply and broadly that his teaching remains to this hour among the most precious possessions of our race.[5]

So far, matters are proceeding as we might have expected from our polemicist. But then, at the precise point when he normally begins to rain down blows on Christianity for spoiling everything again, White suddenly switches to a different tune. Seemingly out of nowhere, and contrary to much of the rest of his grand composition, he begins to sing—in full voice, no less—the praises of the Church:

> With the coming in of Christianity a great new chain of events was set in motion which modified [the Greek] development most profoundly. The influence of Christianity on the healing art was twofold: there was first a blessed impulse—the thought, aspiration, example, ideals, and spirit of Jesus of Nazareth. This spirit, then poured into the world, flowed down through the ages, promoting self-sacrifice for the sick and wretched. Through all those succeeding centuries, even through the rudest, hospitals and infirmaries sprang up along this blessed stream. . . . Vitalized by this stream, all medieval growths of mercy bloomed luxuriantly.[6]

Well, well, well. This is quite the turn. And yet White was genuinely onto something, for the Christian religion was indeed well suited to advancing the cause of biology.

How so?

## Had Not Such Men

The Old Testament—a vast and ancient document considered holy by Jews and Christians alike—is predominantly a mix of religious instruction, historical tales, songs, and poems. Unbeknown to many, however, it also contains some pretty lengthy sections on medical advice. Take this excerpt, for instance, from the book of Leviticus:

> When someone has a burn on their skin and a reddish-white or white spot appears in the raw flesh of the burn, the priest is to examine the spot, and if the hair in it has turned white, and it appears to be more than skin deep, it is a defiling disease that has broken out in the burn. The priest shall pronounce them unclean; it is a defiling skin disease. But if the priest examines it and there is no white hair in the spot and if it is not more than skin deep and has faded, then the priest is to isolate them for seven days. On the seventh day the priest is to examine that person, and if it is spreading in the skin, the priest shall pronounce them unclean; it is a defiling skin disease. If, however, the spot is unchanged and has not spread in the skin but has faded, it is a swelling from the burn, and the priest shall pronounce them clean; it is only a scar from the burn. (Lev. 13:24–28)

Leviticus is around the same age as the Edwin Smith papyrus—and this particular passage goes on to recommend that those who are declared "unclean" should shave off all their hair and wash themselves thoroughly. It seems, then, that there was some sort of early awareness within the Israelite culture of both infection and contamination long before the likes of Hippocrates had arrived on the scene.

The New Testament—which is the defining text for Christendom—also has a surprisingly medical dimension to it. Two of its major books, Luke and Acts, were written by the Apostle Luke—who was, himself, a highly methodical physician. The Apostle Paul, who wrote much of the rest of its content, also turns doctor at one point: he advises a young church leader to "stop drinking only water, and use a little wine because of your stomach and your frequent illnesses" (1 Tim. 5:23).

And, just in case those two revered saints were not exalted enough, then there was always Jesus to look to. After all, Christians worship him as God incarnate, so what he has to say about a subject is of the utmost importance. So, when he told his followers in no uncertain terms to look after the weak, the poor, the lowly, and the sick, one might expect that they would try and take his instruction rather seriously (cf. Matt. 25:31-45).[7]

All of this—ancient Egypt, the Old Testament, the Greeks, the New Testament—bodes very well indeed for the (past) future of medicine. In the fourth century after Christ, much of the world was sitting under the influence of his religion. This new religious empire had doctors among its spiritual leaders, saw looking after the unwell as a holy duty, and had inherited much good practice from its forebears. Perhaps this is why even Andrew Dickson White—who literally wrote the book on the warfare between Christendom and science—could find legitimate grounds for optimism.

And yet, what White was willing to give with one hand, he quickly took back with the other—for, he says, all this promise ended up coming to nothing:

> Had not such men as Thomas Aquinas, Vincent of Beauvais, and Albert the Great been drawn or driven from the paths of science into the dark, tortuous paths of theology, leading no whither,—the world today, at the end of the nineteenth century, would have arrived at the solution of great problems and the enjoyment of great results which will only be reached at the end of the twentieth century, and even in generations more remote. Diseases like typhoid fever, influenza and pulmonary consumption, scarlet fever, diphtheria, pneumonia, and *la grippe*, which now carry off so many most precious lives, would have long since ceased to scourge the world.[8]

Thanks to the "tortuous paths of theology," medicine suffered—and suffered badly. Christianity, whatever Christ might have said, had opposed it, distracted from it, and severely stunted its early growth. The killer diseases of his own era, White said, should have been killed off themselves long ago— and the fact that they hadn't been was the fault of the Church.

This is a radical claim indeed—but he felt he had the evidence to support it.

## The Worst Blow They Ever Received

The first exhibit that White presents is this: Christianity unilaterally banned, under all conceivable circumstances, the internal examination of the dead.

While he admits that this was not a unique position—some preceding cultures had also prohibited the practice—Christendom, he says, took the matter to a whole new level altogether:

> To these arguments [from earlier worldviews] against dissection was now added another—one which may well fill us with amazement. It is the remark of the foremost of recent English philosophical historians, that of all organizations in human history the Church of Rome has caused the greatest spilling of innocent blood.... Strange is it, then, to note that one of the main objections developed in the Middle Ages against anatomical studies was the maxim that "the Church abhors the shedding of blood."[9]

This is polemical writing at its best. White points out what he considers to be an appalling hypocrisy: Catholicism had spilled more blood than any other philosophy had, but had banned dissection on the basis of its bloodiness. Not only was this hypocritical in the extreme, but it was also devastating for medicine:

> On this ground, in 1248, the Council of Le Mans forbade surgery to monks. Many other councils did the same, and at the end of the thirteenth century came the most serious blow of all; for then it was that Pope Boniface VIII, without any of that foresight of consequences which might well have been expected in an infallible teacher, issued a decretal forbidding a practice which had come into use during the Crusades, namely, the separation of the flesh from the bones of the dead ... it soon came to be considered as extending to all dissection, and thereby surgery and medicine were crippled for more than two centuries; it was the worst blow they ever received, for it impressed upon the mind of the Church the belief that all dissection is sacrilege.... So deeply was this idea rooted in the mind of the universal Church that for over a thousand years surgery was considered dishonourable.[10]

White does admit that dissection began to take place in Europe from the thirteenth century onward, but is careful to point out that it only happened in universities "which had become somewhat emancipated from ecclesiastical control." Over time, he says, these less dogmatic institutions grew in confidence, and the landscape gradually changed for the better. By the time we get to the sixteenth century, then, medicine had developed a small but visible chance of recovery:

> Finally came a far greater champion of scientific truth, Andreas Vesalius, founder of the modern science of anatomy. The battle waged by this man is one of the glories of our race. From the outset Vesalius proved himself a master. In the search for real knowledge he risked the most terrible dangers, and especially the charge of sacrilege, founded upon the teachings of the Church for ages.[11]

This hero of White's, Vesalius (1514–1564), quite literally turned dissection into an art form. He oversaw the production of the most wonderfully detailed woodcuts, each portraying the discoveries he had made when opening up his subjects' bodies. It was a huge leap forward for human anatomy, and even now his diagrams still populate many textbooks.

His great service to science cost him dearly, however—for, according to White, despite many years of "braving the fires of the Inquisition," Vesalius was eventually brought down. The Church—one way or the other—managed to end his unholy work permanently:

> Vesalius was charged with dissecting a living man, and, either from direct persecution, as the great majority of authors assert, or from indirect influences, as the recent apologists for Philip II admit, he became a wanderer: on a pilgrimage to the Holy Land, apparently undertaken to atone for his sin, he was shipwrecked, and in the prime of his life and strength he was lost to the world.[12]

Thus it was that the organization which "abhors the shedding of blood" deliberately brought about century after century of unnecessary deaths. What else may have been learned from Vesalius, and others before and after him, if only the Church had behaved differently? How many patients might have been saved?

Still, its ban on dissection and its persecution of anatomists were not the only failings of the Church, White says. In another shocking move, its irrational dogma gave rise to something very nearly as bad as pointless death.

Namely, pointless pain.

## Contrary to Holy Writ

In *Physicians, Plagues and Progress,* Oxford historian of science Allan Chapman (whom we met in the last chapter) manages to cover pretty much the entire history of medicine—from ancient Egyptian operations to the Ebola epidemic of 2015. During nearly all of this vast timespan, despite all the changes and development, he says, medical professionals have come face to face with one terrifying constant:

> Whether the surgeon treated the wounded in Pharaoh's army, as in the Edwin Smith Papyrus of c. 1500 BC, whether he were Galen with his gladiators, John of Ardene with his medieval knights, John Hunter or Sir Astley Cooper at Guy's and St Thomas's Hospitals in 1830, they all had one thing in common: the necessary infliction of appalling suffering upon their patients.[13]

There had been numerous attempts, of course, to limit this "appalling suffering" over the years—not least by using copious amounts of alcohol. In the end, however, a consensus developed. The best thing to do, it was decided, was to forget the idea of preventing pain, and to focus on something else instead: speed.

A lightning-quick cut, followed by a lightning-quick clean (in some cases) and lightning-quick stitching or cauterizing became the new goal for the best of the best. As techniques improved, some surgeons in the early nineteenth century could take off a limb and sort out the ensuing mess in less than a minute.

One of the biggest names in the business was Scotsman Robert Liston (1794–1847)—a man so celebrated that he was still earning praise from the *British Medical Journal* in 1908, more than half a century after his death:

> He was the foremost operator in London, his powerful "shoulder of mutton fist" and his wonderful skill with the knife enabling him to deal successfully with cases which other surgeons preferred to leave alone. It is told of him that, when he amputated, the gleam of his knife was followed so instantaneously by the sound of the saw as to make the two actions appear almost simultaneous.[14]

Still, as fast as some of these operations might well have been, they were hardly a pleasant experience. Screams, convulsions, and patients thrashing around wildly were everyday occurrences in the life of a surgeon.

But then, right in the middle of Liston's heyday, a new technology emerged in America which promised a previously unimaginable scenario—surgery without any pain. An innovator as well as an outstanding practitioner, Liston was one of the first to make use of this almost magical ether in 1846. Despite his own reference to the substance as a "Yankee Dodge," all went successfully. The game had changed—forever.

James Young Simpson (1811–1870) of Edinburgh, another talented surgeon, was fascinated. After finding that ether was somewhat unpredictable in its effects, he sought out an alternative of his own. Arriving at the very new chloroform, he found it worked well enough on himself and his friends for him to start using it in his day job: obstetrics. Wonderfully, women would now be able to have some sort of pain relief during childbirth; a curse as old as history itself was finally being lifted—by the strong and dependable arms of empirical science.

And yet, as White has already warned us, whenever progress is on offer, theology will find a way to stand against it. The glorious triumph of anesthetic, *Warfare* recounts, was no exception:

> From pulpit after pulpit Simpson's use of chloroform was denounced as impious and contrary to Holy Writ; texts were cited abundantly, the ordinary declaration being that to use chloroform was "to avoid one part of the primeval curse on woman."

Simpson had seen this coming—for, incredibly, even some of his own pregnant patients had warned him off. He wrote that some had told him "an immunity from pain during [childbirth] was contrary to religion and the express commands of Scripture."[15]

Thinking this through, Simpson decided to push back—and, in 1847, he published his *Answer to the Religious Objections Advanced Against the Employment of Anaesthetic Agents in Midwifery and Surgery*. In it, the Scotsman came up with an ingenious riposte for his opponents:

> I allude to . . . the first surgical operation ever performed on man, which is contained in Genesis ii. 21:—"And the Lord God caused a deep sleep to fall upon Adam; and he slept; and he took one of his ribs, and closed up the flesh instead thereof." In this remarkable verse the whole process of a surgical operation is briefly detailed. But the passage is principally striking, as

affording evidence of our Creator himself using means to save poor human nature from the unnecessary endurance of physical pain.[16]

Clever stuff.

Until now, the reader may note, we have not heard anything on the subject from a certain John William Draper. This is, admittedly, partially due to the fact that *Conflict* is significantly shorter than its cousin, and therefore covers less material—but the Church's sheer callousness when it came to pain relief was so galling that Draper also decided to make room for it:

> When the great American discovery of anæsthetics was applied in obstetrical cases, it was discouraged, not so much for physiological reasons, as under the pretense that it was an impious attempt to escape from the curse denounced against all women in Genesis iii. 16.[17]

And, while on medical matters, Draper dropped in another pious pearler:

> When the Mohammedan discovery of inoculation was brought from Constantinople in 1721, by Lady Mary Wortley Montagu, it was so strenuously resisted by the clergy, that nothing short of its adoption by the royal family of England brought it into use.[18]

This additional point had not escaped White, either:

> English theologians were most loudly represented by the Rev. Edward Massey, who in 1772 preached and published a sermon entitled *The Dangerous and Sinful Practice of Inoculation*. In this he declared that Job's distemper was probably confluent smallpox; that he had been inoculated doubtless by the devil; that diseases are sent by Providence for the punishment of sin; and that the proposed attempt to prevent them is "a diabolical operation." . . . This struggle went on for thirty years.[19]

So, the Church had banned dissection and autopsies, opposed anesthetic, even for women in labor, and then thrown the Bible at inoculation and vaccination. So much for building on early promise. So much for looking after the sick.

And Draper and White were not done yet.

## Monstrosities of Error

The various frictions mentioned thus far, *Conflict* and *Warfare* were both keen to point out, were merely symptoms of an underlying illness. The real issue was not opening cadavers, or administering pain relief, or preventing disease; it was, instead, the fundamental opposition of Christianity to the concept of medicine as a whole.

Here, for instance, is Draper's take: "It had always been the policy of the Church to discourage the physician and his art; he interfered too much with the gifts and profits of the shrines."[20] And so "for the prevention of diseases, prayers were put up in the churches, but no sanitary measures were resorted to. From cities reeking with putrefying filth it was thought that the plague might be stayed by the prayers of the priests."[21] White agrees: " . . . we find cropping out every where the feeling that, since supernatural means are so abundant, there is something irreligious in seeking cure by natural means: ever and anon we have appeals to Scripture."[22]

The grand result of all this supernatural irrationality, they allege, was a descent into farce. Draper—rather humorously, it must be said—explains that demand for holy relics got so big that the Church was forced to increase its supply:

> There were several abbeys that possessed our Savior's crown of thorns. Eleven had the lance that had pierced his side. If any person was adventurous enough to suggest that these could not all be authentic, he would have been denounced as an atheist. During the holy wars the Templar-Knights had driven a profitable commerce by bringing from Jerusalem to the Crusading armies bottles of the milk of the Blessed Virgin . . . none of these impostures surpassed in audacity that offered by a monastery in Jerusalem, which presented to the beholder one of the fingers of the Holy Ghost![23]

He then comments that, having finally seen scientific sense: "modern society has silently rendered its verdict on these scandalous objects."[24]

*Warfare* has a fair old go too, making sport of yet more religious silliness:

> Closely connected with these methods of thought was the doctrine of signatures. It was reasoned that the Almighty must have set his sign upon the various means of curing disease which he has provided: hence it was

held that bloodroot, on account of its red juice, is good for the blood . . . eye-bright, being marked with a spot like an eye, cures diseases of the eyes . . . bu-gloss, resembling a snake's head, cures snakebite . . . bear's grease, being taken from an animal thickly covered with hair, is recommended to persons fearing baldness.[25]

In other words, had a medieval monk wanted to mend a maddening migraine, he would have foraged for a divinely designed object in nature which physically resembled the problem. James Hannam, historian and author, takes up the story:

> No inspired guesswork is required. The [walnut] has a hard outer shell which, when cracked open, reveals a soft craggy interior divided into hemispheres. The similarity between the edible part of a walnut and the brain must have struck herbalists as undeniable. This signature was nature's way, or God's way for a Christian healer, of showing man which plant to use to treat a head complaint.[26]

This feels like quite the backward step from the careful detail of the Edwin Smith Papyrus. Here we are, three thousand years on from that medical masterpiece, and the Church has given people walnuts for brains instead.

This rather depressing collapse has caused many a lament. Walter Clyde Curry (1887–1967), poet and historical writer, famously said that medieval medicine was chock-full of "monstrosities of error."[27] His contemporary in the field, heavyweight thinker Charles Singer (1876–1960), was equally damning: "Surveying the mass of folly and credulity . . . it may be asked, is there any rational element here? The answer, of course, is very little."[28]

And so—despite its wonderful early start with the Egyptians, the Old Testament, the Hippocratic Oath, patient histories, Aristotle's evidence-based approach, Greek dissection, and the bright hope of Jesus's command to care for the sick—late medieval biology, under Christianity, seems to offer "very little."

Perhaps, without Christendom's centuries-long stranglehold on science, things would have been different. Perhaps we would now know why we pick our noses, or why we yawn, or even what consciousness is. Perhaps, in fact, the strange events surrounding Gloria Ramirez might never have happened in the first place—for, without the constant interference of dogma, we might already have achieved that oh-so-precious goal: a cure for cancer.

The case against the Church here seems overwhelming. We must remember, however, that it also seemed overwhelming as regards the flat Earth—yet that particular house turned out to be made of straw. What will happen, then, if we start huffing and puffing strongly enough at the Christianity-messed-up-medicine dwelling?

Will it fall down, too?

## I Was Jolted to Discover I Was Wrong

Well, if straw does eventually emerge as the dominant material once more, we would have to conclude it had been put there by some rather reputable builders—for Draper and White enjoy the backing of quite a few famous names. Here, for example, is an excerpt from the highly esteemed British Library's entry on Vesalius: "For 1000 years after Galan's [sic] death almost no original anatomical inquiries were performed, mainly because the Church was against the dissection of human bodies."[29]

*Conflict* and *Warfare*'s chorus line is also repeated in the 2009 edited volume *Surgery: Basic Science and Clinical Evidence*, which was "heartily recommended" by none other than the *Journal of the American Medical Association*:

> By the end of the Middle Ages [AD 1500 onward], it had become apparent to physicians that further progress in the knowledge of medicine, specifically surgery, could not be attained unless scientific studies of human anatomy were made. By that time, the church's ban on human vivisection was showing signs of weakness.[30]

The British Broadcasting Corporation is also with Draper and White. On a revision page aimed at the quarter of a million high school students who take the history GCSE each year, it wrote: "Causes of medical stagnation in the Middle Ages included the forbidding by the Church of dissection, and its encouragement of prayer (and superstition)."[31]

Even the US Senate has toed the line. Senator Arlen Specter, during a 2005 debate about stem cell research, warned those present to learn the lessons of the past—for they must make sure, he urged, that Christian dogma did not hold medicine back as it had before: "Pope Boniface VII [sic] banned the practice of cadaver dissection in the 1200s. This stopped the practice for over 300 years and greatly slowed the accumulation of education regarding human anatomy."[32]

This is a weighty crowd. Could the British Library, the American Medical Association, the BBC, and the Senate really all be wrong?

Surely not.

And yet, in a 2009 article written by the journalist Christopher Howse for the *Telegraph*, there is a hint that they just might be:

> I thought I was fairly immune to popular myths and vulgar errors about science and religion. Hardly anyone believed in a flat Earth in the Middle Ages, I knew. . . . So I was jolted to discover I was wrong in supposing that the medieval Church forbade human dissection.[33]

Hmm. Can we get to the bottom of this?

## Not Only Tolerated but Encouraged

Let's cut to the chase: the Church did not ban dissection. Likewise, it did not ban autopsies, or anesthetic, or inoculation. We can go further: Christendom did not even oppose these things, either. Instead, Christians were often at the forefront of both their development and their use. All of which means that the story, as Draper and White and many exalted others tell it, is not far from being completely backward.

Let's tell it again, then—but this time, we shall do it the right way around.

Firstly, the common notion that the Greeks and Romans were dissecting here, there, and everywhere is simply false. Other than the aforementioned Herophilus and Erasistratus, the somewhat surprising fact is that no one in Greece or Rome carried out medical dissection, either before them or after them. This is why the classicist and expert in ancient medicine Heinrich von Staden decided to ask, in what would become a groundbreaking paper: "What had rendered the practice impossible for so long [beforehand]? What rendered it impossible again for more than 15 centuries after Herophilus and Erasistratus?"[34]

Digging through key documents covering many centuries and many cultures, von Staden concluded that it was simply the position of nearly everyone nearly all the time across the ancient world that dead bodies were not to be messed with. This view was a constant, regardless of religion, language, mythology, and level of scientific sophistication. Even societies in which the dead were embalmed treated those who actually did the deed as highly circumspect and, in one way or another, "unclean."

In the light of von Staden's study, White's earlier analysis can be flipped on its head. For, when dissection eventually did resurface in the twelfth and thirteenth centuries—which it did, and as a learning tool, no less—it happened under the auspices of Catholicism, in its universities and schools, where it was given both its legitimacy and its sponsorship by the leaders of the Church.

Indeed, should we permit ourselves the same level of polemic license as White did, we could put it this way: it was the ancients (Greeks included) who placed a ban on dissection, and the Church that legalized it. We could say that it was Christendom that finally freed medicine from this irrational dogma. Katharine Park, the Harvard professor who brought our last chapter to a close, shows just how wrong Draper and White really were: "Most medieval church authorities not only tolerated but encouraged the opening and dismemberment of human corpses."[35]

So, what were the two of them up to? How did they end up asserting the very opposite of the truth?

### Never, Ever, Anywhere

Well, we should perhaps begin with one of White's chief pieces of evidence. His thirteenth-century edict that supposedly did all the damage—"the Church abhors the shedding of blood"—turns out, upon further investigation, to be a rather problematic one.

As far as anyone can tell, it is first found not in the 1200s, but in the 1700s. French surgeon François Quesnay (1692–1774) had read the work of an earlier historian, Étienne Pasquier (1529–1615), and had decided to borrow some of it for his own writings. He translated one of Pasquier's own phrases about the Church and blood into Latin, placed the result between quotation marks, and stuck it into his text. One can only suppose that others saw this and assumed that this entirely new expression actually had much earlier origins—and, just like that, it seems a myth was born.[36]

White, for what it is worth, does not quote Quesnay as his source—instead, he references other historians sitting in the gap between him and the Frenchman. Perhaps the slogan was so succinct, so believable, and so useful that it ended up spreading far and wide fairly quickly, despite its lack of genuine historicity—we cannot be sure. What we do know is that in the 1960s, historian C. H. Talbot tried to chase it down, and found that all roads ended with Quesnay. He labeled it "a literary ghost" and, having searched the medieval literature accordingly, confirmed that "no earlier source for this sentence can be found."[37]

Then, in 1978, classicist Darrell Amundsen set out on a similar quest. After wading through every relevant Church document from the Middle Ages and failing to find the phrase anywhere, he commented that:

> It is frequently claimed that the Church forbade the practice of surgery to all clerics on the ground that *Ecclesia abhorret a sanguine*, that is, "The Church abhors the shedding of blood." This maxim is sometimes attributed to canon 18 of Lateran IV although usually to the Council of Tours, which took no action on the question of the practice of surgery.[38]

White's second piece of evidence is also misleading. He claims that Pope Boniface's decretal about not boiling flesh from the bones of dead crusaders "soon came to be considered as extending to all dissection," but that is a gross exaggeration. Yes, a small number of folk misunderstood it, took his instruction too far, and backed off somewhat—yet they were in the tiny minority. The vast majority carried on with their practices, unperturbed. Andrew Cunningham, in his study *The Anatomist Anatomis'd*, sums up the situation rather bluntly:

> As a life-long evangelical atheist I certainly hold no brief for the Catholic Church. Nevertheless the fact is that the Catholic Church has never been opposed to the practice of anatomy, whether for post-mortem, demonstration, teaching or research purposes. Never, ever, anywhere.[39]

Park agrees: "I know of no case in which an anatomist was ever prosecuted for dissecting a human cadaver and no case in which the church ever rejected a request for a dispensation to dissect."[40] And, while she is on the topic, Park takes White to task on another of his assertions: "There is no convincing evidence that Vesalius ran afoul of church authorities two hundred years later."[41]

It is now Cunningham's turn to agree with Park:

> Although White claims that Vesalius was attacked using "weapons theologic" . . . he presents no evidence or referencing for such claims. It would be

difficult to do so, since they are figments of White's imagination. It must be remembered that this is a work of propaganda, and the truth rarely bothers propagandists.[42]

White, then, has misled people on dissection, on Church history, and on the life and fate of Vesalius. So believable was his account that he has duped some of the most eminent authorities of our own time, who continue to propagate his exaggerations and untruths. The late and masterly historian of science Colin Russell (1928–2013) provides a fascinating insight into the lasting effect of *Warfare's* folly:

> I had a research student who looked at this as part of a bigger task and, to cut a long story short, he found that all the stories to this effect, which are in all the modern textbooks on the history of medicine, are actually almost baseless; but not quite, because he traced them down in a sort of family tree to just one source, and that one source was A.D. White.[43]

Once again, then, we have a clear-cut example of a world largely fooled. The British Library was sucked in. The American Medical Association was tricked. The BBC were hoodwinked.

And Senator Arlen Specter—rather fittingly, perhaps—had referenced a literary ghost.

## An Untested Novelty

As it happens, Draper and White did not just get dissection horribly wrong, but anesthetic and inoculation too. While it is true that Simpson the obstetrician had felt the need to defend his use of chloroform against religious zealots—even going so far as to write a pre-emptive booklet on the topic—the big showdown he had anticipated never actually materialized.

White, in *Warfare*, had us believe that Simpson was preached against all over the place, and that his life-improving actions had been denounced, universally, as ungodly—but this is just not true. In actual fact, the vast majority of those who did oppose pain relief during childbirth were not priests at all—they were doctors.

According to many of the professionals at the time, pain was a necessary part of labor—for without it, they argued, both mother and baby might come to serious harm. These physicians feared that a sleeping woman would not be able to push at the right moments, and that the result could be disastrous for her and her child. After Simpson was able to demonstrate that such worries were wholly unfounded, however, their complaints quietly melted away. And that was that.[44]

It is a similar story with inoculation. Draper wrote that it had been "strenuously resisted by the clergy"—but it hadn't. There certainly was strenuous resistance, but it was not from the clergy. Boston, for example, had been hit by smallpox in the 1720s, and had faced a decision on whether or not to gamble on the new technique. Here is what happened:

> Boston ministers, guided by [Reverend] Cotton Mather, for the most part stood in favor of extending the experiment in inoculation. . . . Most physicians, on the other hand, led by Dr. William Douglass, the only practitioner in town with a medical degree, opposed inoculation as an untested novelty, attacked the clergymen who fostered it, and called upon the town selectmen to halt the experiment.[45]

In this landmark New England case, then, we find that Draper's broad assessment is both upside-down and inside-out. Historian Maxine Van de Wetering has called his bluff: it was the medicos pleading for caution, and the preachers championing the "untested novelty."

The debate, by the way, became rather intense: at the height of the argument, Mather actually had a bomb thrown into his house. Fortunately, it failed to go off—but a note was subsequently found attached to it which read "you dog, dam you! I'l inoculate you with this; with a pox to you."[46]

Ultimately, the statistics came down on the side of Mather, and inoculation won out. Did some individual firebrands denounce it in some sermons in some churches either side of the Atlantic? Yes. Did the Church, as a unified body, oppose it? Did "Christendom"? No. Instead, some of its most influential members, Mather included, were the ones who gave it life.

## It Just Kills Me

The big medical picture of *Conflict* and *Warfare*—complete with carefully painted details of dissection, anesthetic, and inoculation—is suddenly not looking all that great. It is a very poor representation of reality; it is an unreliable guide. When Draper wrote that "It had always been the policy of the Church to discourage the physician and his art" he could hardly have been more wrong—and historians have been demonstrating as much for decades.

Nearly a century and a half later, though, the Draper–White portrait is still being faithfully reproduced by those supposedly in the know. Here, for instance, is a paper in the peer-reviewed medical journal *Anatomy & Cell Biology*, published as recently as September 2015:

> The flickering light of human dissection was completely snuffed out with the burning of Alexandria in 389 AD. Following widespread introduction of Christianity in Europe during the Middle Ages, the development of rational thought and investigation was paralysed by the church authorities. . . . During this period, human dissection was considered to be blasphemous and so was prohibited. For hundreds of years, the European world valued the sanctity of the church more than scientific quest. . . . One of the significant proscriptions that Pope Alexander III enunciated at the Council of Tours in 1163 was . . . named as "Ecclesia abhorret a sanguine" meaning "The church abhors blood."[47]

Almost every statement in the passage above is wrong—and yet the article has over a hundred footnotes in total, and there are plenty even in the short section we have quoted. Many of these cite other peer-reviewed journals, including the enormously prestigious *Journal of the American Medical Association*. Draper and White, it would seem, are everywhere. It is little wonder, then, that Park seems almost ready to give up:

> Every time I read something in The New York Times that Leonardo da Vinci had to hide the fact that he was doing dissection, and every time I listen to a tour guide in Italy tell these stories, it just kills me. I don't know how to get rid of this myth.[48]

### Relics and Bear's Grease

Perhaps, however, we should not entirely dismiss *Conflict* and *Warfare* and their despair at the dreadful state of medieval medicine. After all, if we do a little checking, we can quickly find that the doctrine of signatures was a real thing. The same can be said of the sin theory of disease, and of blood-letting, and purges, and potions, and incantations, and pious visits to various shrines.

We must be careful, then, not to throw out some babies with their conflict thesis bathwater. After all, when a historian as distinguished as Charles Singer can survey pretty much the entire field of medicine in the Middle Ages and conclude that it "lacked any rational element,"[49] we have to take him seriously—don't we?

Well yes, we do—but that does not mean that we cannot also question such received wisdom. And, in the years since Singer hung up his typewriter, plenty have done so. What they have found has led to what could probably be called a mini-revolution in the discipline: for it seems that the Dark Ages were not all that dark after all: "it becomes clear that medieval intellectuals were not only rational; they were *hyper*rational—to them, logic was . . . the single most important tool for the investigation of the natural world."[50] So says one of the new breed of medievalists, Richard Raiswell. Yet this is hardly the image we were left with by Draper, and White, and Singer, and Specter, and the British Library, and all those others. Were the Dark Ages really so badly named? Could the folk who lived back then—committed as they were to the healing powers of questionable relics and bear's grease—truly be called hyperrational with a straight face?

Let's find out.

# 5

# Tales from the Gap

## Top Tens

Everyone loves a top ten. Nowadays, no sport is free from a discussion about its ten best ever players, or ten best ever scores, or ten best ever teams. No superhero universe is complete without a rundown of its ten biggest, or fastest, or most powerful characters. The website TheTopTens currently has 187,481 lists on it—including those on musicians, actors, and digital camera brands. There are even a number of lists of what are considered to be the best top ten lists. And yes, at least one joker has ranked those, too—forming a "Top Ten List of Top Ten Top Ten Lists."[1]

The UK's *Guardian* newspaper has been in on this act for a while, with a regular slot called "The 10 Best . . ." In April 2010 it was the turn of the mathematicians to enjoy the spotlight. Sure enough, some of the big hitters were selected—Pythagoras, Euler, Gauss—but one of the other names to make the grade might be slightly less well known: that of Hypatia of Alexandria.[2]

Hypatia was a remarkable woman. She was a renowned scholar and a teacher, and she (along with her father, Theon) presided over the school and library of the *Serapeum*, a temple in the cosmopolitan North African city. Her reputation was superb; she attracted the brightest and best (and richest) of students from lands afar. These learners would sit at her feet in wonder as she unlocked Euclid's geometry, Ptolemy's astronomy, or Diophantus's algebra in her famously lucid style.

We don't know for sure which decade Hypatia was born in—but we do know precisely when she died. For, as the *Guardian* explains, in the spring of AD 415 "she was murdered by a Christian mob who stripped her naked, peeled away her flesh with broken pottery and ripped apart her limbs."[3]

Oh.

## Philosophy and Bigotry

The shockingly brutal murder of Hypatia has, quite understandably, made her a subject of fascination for historians, authors, and artists alike. The flamboyant Englishman Edward Gibbon (1727–1794) was one of many who have dwelt on her story—it is a key event in his paradigm-shifting *The History of the Decline and Fall of the Roman Empire* (1788):

> In the bloom of beauty, and in the maturity of wisdom, the modest maid refused her lovers and instructed her disciples . . . and Cyril [Archbishop of Alexandria] beheld with a jealous eye the gorgeous train of horses and slaves who crowded the door of her academy. . . . On a fatal day, in the holy season of Lent, Hypatia was torn from her chariot, stripped naked, dragged to the church, and inhumanly butchered by the hands of Peter the reader and a

troop of savage and merciless fanatics ... the murder of Hypatia has imprinted an indelible stain on the character and religion of Cyril of Alexandria.[4]

More recently, the acclaimed writer/director/composer Alejandro Amenábar has brought Hypatia back to life—in the 2009 film *Agora*. Praised for its lurid and gritty depiction of end-of-empire Alexandria, *Agora* has warring factions of Christians, pagans, and Jews taking assorted pot-shots at one another. At its center is the *Serapeum*—in which are housed precious scrolls of wisdom gathered together, over the centuries, from every corner of the ancient world.

Early on in the film, the black-robed and crazed Christians—who despise Hypatia and all that she stands for—viciously attack the library, smashing its statues, and burning its manuscripts. Hypatia, shaken but undeterred, continues to teach and study for another two decades—but the viewer can sense it will all end in tragedy.

In the period between this first attack and her eventual death, *Agora* treats us to the stark contrast between Hypatia and Christianity. While the Christians attack and kill Jews, Hypatia makes the astonishing discovery that the Earth orbits the Sun, and that it does so in an ellipse. While the Christians scheme against the pagans, Hypatia tests out her ideas about the Earth's motion by performing inventive experiments on falling objects. She reminds the archbishop that he, as a man of faith, cannot question his beliefs—whereas she, as a philosopher, is duty bound to.

Ultimately, the inevitable arrives. Hypatia's atheistic science and her level of influence as a powerful woman prove far too offensive to the Church and, on the orders of Saint Cyril, she is killed. Her extraordinary mind—just like her extraordinary library—is lost to the world, forever.

To describe all this as a haunting injustice seems wholly appropriate. In *The Darkening Age: The Christian Destruction of the Classical World*, Cambridge classicist-turned-journalist Catherine Nixey rages at the conduct of the early Church in first destroying the *Serapeum*:

Nothing was left. Christians took apart the very temple's stones. ... The tens of thousands of books, the remnants of the greatest library in the world, were all lost. ... As the modern scholar, Lucian Canfora, observed: "the burning of books was part of the advent and imposition of Christianity."[5]

and then, years later, destroying its mistress:

[they] dragged Alexandria's greatest living mathematician through the streets to a church. ... Some say that, while she still gasped for breath, they gouged

out her eyes. Once she was dead, they tore her body into pieces and threw what was left of the "luminous child of reason" onto a pyre and burned her.[6]

This is, perhaps, the perfect example of the conflict thesis in action: it is religion throwing its full weight at science. It is dogma murdering reason. And John William Draper, as we might expect, was appalled by it:

> [The Christian Emperor Theodosius I] dispatched a rescript to Alexandria, enjoining the bishop, Theophilus, to destroy the Serapion; and the great library, which had been collected by the Ptolemies, and had escaped the fire of Julius Caesar, was by that fanatic dispersed.
>
> Hypatia and Cyril! Philosophy and bigotry. They cannot exist together. So Cyril felt, and on that feeling he acted. . . . For this frightful crime Cyril was never called to account. It seemed to be admitted that the end sanctified the means. So ended Greek philosophy in Alexandria, so came to an untimely close the learning that the Ptolemies had done so much to promote.
>
> The fate of Hypatia was a warning to all who would cultivate profane knowledge.
>
> Henceforth there was to be no freedom for human thought.[7]

The Church, then, had nailed its colors to the mast—their empire was to be one of doctrine, not of freethought. No longer would Greco–Roman science and philosophy lead the way, for their new order would be one of blind and subservient faith instead. And, perhaps conveniently, Hypatia's bloody death would serve as a warning to anyone who was not quite ready to comply.

## Contempt upon All Investigators

Draper, Nixey, and Gibbon are not the only ones to see Hypatia's gruesome end as a bit of a watershed. David C. Lindberg remarks that there are many for whom "Hypatia's murder marked the 'death-blow' to ancient science and philosophy," and gives some examples:

> The distinguished historian of science B. L. Van der Waerden claims that "[a]fter Hypatia, Alexandrian mathematics came to an end"; in his study of ancient science, Martin Bernal uses Hypatia's death to mark "the beginning of the Christian Dark Ages."[8]

Ah, yes—the "Dark Ages."

In his highly successful 2002 book *The Closing of the Western Mind*, Charles Freeman explains how Christendom effectively called time on the Greeks and the Romans, replacing their science and rationality with an unquestioning trust in Scripture and the Church. Freeman opens his work with two contrasting quotations to illustrate his point. Firstly, here is the Greek literary genius Euripides (c. 480–406 BC):

> Blessed is he who learns how to engage in inquiry, with no impulse to harm his countrymen or to pursue wrongful actions, but perceives the order of immortal and ageless nature, how it is structured.[9]

And, secondly, here is the esteemed Church Father, Saint Augustine (AD 354–430):

> There is another form of temptation, even more fraught with danger. This is the disease of curiosity. . . . It is this which drives us to try and discover the secrets of nature, those secrets which are beyond our understanding, which can avail us nothing and which man should not wish to learn.[10]

It is clear, Freeman argues, where the blame for the Dark Ages lies—and it isn't with the Greeks.

He is far from being the first to say so—Draper, in *Conflict*, can barely hide his fury at Augustine:

> No one did more than this Father to bring science and religion into antagonism; it was mainly he who diverted the Bible from its true office—a guide to purity of life—and placed it in the perilous position of being the arbiter of human knowledge, an audacious tyranny over the mind of man. The example once set, there was no want of followers; the works of the great Greek philosophers were stigmatized as profane; the transcendently glorious achievements of the Museum of Alexandria were hidden from sight by a cloud of ignorance, mysticism, and unintelligible jargon . . .[11]

*Warfare*, too, gets in on the game—initially by praising the ancients:

> [the Greeks and Romans] gave scientific freedom. . . . This legacy of belief in science . . . was especially received by the school of Alexandria, and above all by Archimedes, who began, just before the Christian era, to open new

paths through the great field of the inductive sciences by observation, comparison, and experiment.[12]

and then by bashing Christendom for ruining everything:

The establishment of Christianity, beginning a new evolution of theology, arrested the normal development of the physical sciences for over fifteen hundred years . . . there was created an atmosphere in which the germs of physical science could hardly grow . . . the greatest thinkers in the Church generally poured contempt upon all investigators into a science of Nature.[13]

Here, then, are the Dark Ages. No science, no progress, no wisdom, no quality of life—and, Draper says, no hope either:

If by chance a passing interest was taken in some astronomical question, it was at once settled by a reference to such authorities as the writings of Augustine or Lactantius, not by an appeal to the phenomena of the heavens. So great was the preference given to sacred over profane learning that Christianity had been in existence fifteen hundred years, and had not produced a single astronomer.[14]

Freeman, when considering a subtitle for his all-encompassing account of the Church's deadly assault on rational thought, landed upon *The Rise of Faith and the Fall of Reason*.

It says it all, really.

## Of Popes and Unicorns

Ditching science comes at a cost. Medieval beliefs about the world, according to both Draper and White, soon became nonsensical—for minds were being fueled mostly by drivel. Here is *Warfare*:

The most careful inductions from ascertained facts were regarded as wretchedly fallible when compared with any view of nature whatever given or even hinted at in any poem, chronicle, code, apologue, myth, legend, allegory, letter, or discourse of any sort which had happened to be preserved in the literature which had come to be held as sacred.[15]

By rejecting philosophy and the physical sciences, the Church had opened up the intellectual back door to "sacred" silliness. Anyone claiming their ideas were somehow "holy" was given an audience. It was a free-for-all, says White:

> The great work of Aristotle was under eclipse . . . in place of it [Christians] developed the *Physiologus* and the Bestiaries, mingling scriptural statements, legends of the saints, and fanciful inventions with pious intent and childlike simplicity . . . these remained the principal source of thought on animated Nature for over a thousand years.[16]

The "*Physiologus* and the Bestiaries" were fantastical zoological catalogues of God's remarkable handiwork—and no scientific study or evidence, White says, was needed in their composition. He gives some of the laughable results:

> Neglecting the wonders which the dissection of the commonest animals would have afforded them, these naturalists attempted to throw light into Nature by ingenious use of scriptural texts, by research among the lives of the saints, and by the plentiful application of metaphysics. Hence even such strong men as St. Isidore of Seville treasured up accounts of the unicorn and dragons mentioned in the Scriptures.[17]

Oh dear. Archbishop Isidore (c. 560–636) was considered by his contemporaries in the Church to be a God-anointed genius—to be "Solomon revived," no less—and yet, without hint of a blush, he firmly believed in unicorns.[18] And he was not the only one to do so—for here, once more, is our old friend of flat Earth fame, Cosmas Indicopleustes:

> This animal is called the unicorn, but I cannot say that I have seen him. . . . When he finds himself pursued by many hunters and on the point of being caught, he springs up to the top of some precipice whence he throws himself down and in the descent turns a somersault so that the horn sustains all the shock of the fall, and he escapes unhurt. And scripture in like manner speaks concerning him, saying, *Save me from the mouth of lions, and my humility from the horns of unicorns* . . . thus bearing complete testimony to the strength, audacity, and glory of the animal.[19]

Incredibly, White says, the Church was still trotting out anti-scientific tales like this as late as the 1600s. He quotes the French monk Eugene Roger, who reported that a dragon-like basilisk "appeared in Rome and killed many people by merely looking at them; but the Pope destroyed it with his prayers and the sign of the cross."[20]

Draper, not to be outdone, includes his own a story of a medieval pope saving the day:

> An illiterate condition everywhere prevailing, gave opportunity for the development of superstition . . . when Halley's comet came, in 1456, so tremendous was its apparition that it was necessary for the pope himself to interfere. He exorcised and expelled it from the skies. It slunk away into the abysses of space, terror-stricken by the maledictions of Calixtus III., and did not venture back for seventy-five years![21]

This, then, is the fruit of twelve hundred years of Christianity: fairy tales of popes and unicorns. The Church, in choosing to ban Hypatia's math, Plato's philosophy, and Aristotle's physics, had deliberately brought about the greatest brain-drain of all time. And, when we do a little more digging, it looks like much of this stupefying silliness was shooed in by a smattering of sanctified statements from some supposedly sacred saints.

## Laughed at by a Schoolboy

Some time around AD 55, St. Paul wrote the following to a young church in Corinth, Greece:

> Where is the wise person? Where is the teacher of the law? Where is the philosopher of this age? Has not God made foolish the wisdom of the world? . . . Greeks look for wisdom, but we preach Christ crucified. . . . For the foolishness of God is wiser than human wisdom. (1 Cor. 1:20–26)

Paul, as it happens, is not the only important Christian thinker to speak strongly on the topic of philosophy. The hugely influential Tertullian (c. AD 155–220), for example, is well known for writing aggressively against pagan thought, and for encouraging his fellow believers to shut their eyes and ears to it. Historian Winston Black explains that Tertullian:

> associated the study of Greek science and philosophy with the worship of demons, and in this context uttered his most famous phrase: "What indeed

has Athens to do with Jerusalem?" By this rhetorical question, he meant the teachings of ancient Greek philosophers (Athens) have nothing to do with Christ and his teachings (Jerusalem).[22]

And, of course, there is Augustine.

It is difficult to overstate the impact that this North African bishop had on the Christian world—for he flourished at a time when his religion was in the ascendency, and his extensive writings became both widespread and deeply treasured. Both Draper and White name Augustine as the person most responsible for the Christian Dark Ages—in fact, White mentions him on more than eighty separate occasions in *Warfare*. Here is a typical example:

> St. Augustine, preparing his *Commentary on the Book of Genesis*, laid down in one famous sentence the law which has lasted in the Church until our own time: "Nothing is to be accepted save on the authority of Scripture, since greater is that authority than all the powers of the human mind."[23]

This holy law can lead only to irrationality, White laments—and, to make his point, he presents an extraordinary pair of claims he has found in Augustine's masterpiece, *City of God*:

> St. Augustine was certainly one of the strongest minds in the early Church, and yet we find him mentioning, with much seriousness, a story that sundry innkeepers of his time put a drug into cheese which metamorphosed travelers into domestic animals, and asserting that the peacock is so favored by the Almighty that its flesh will not decay.[24]

Yes, that really does say what it appears to say. Augustine, having purposefully taken leave of his senses because of his faith, is now teaching his flock that drugged cheese can turn men into donkeys, and that peacock meat—for God so loved the bird—never rots. White comments that, in his own more scientific age, such ideas would even be "laughed at by a schoolboy."[25]

What chance could the poor passengers adrift in the Middle Ages possibly have, then, when their assorted captains—Paul, Tertullian, Augustine, and more—were consistently rejecting wisdom and teaching nonsense? It is little wonder that the term "medieval" has become synonymous with "backward."

And not only "backward." Displaying extraordinary flexibility, "medieval" can also stand in for "cruel," or "filthy," or "painful," or even "evil." For, as Draper and White and Nixey and Freeman and Gibbon and others have already reported, the Dark Ages were a time of fecklessness, of hopelessness, and of desolation; of darkness, of sickness, and of misery.

Even some of their own said so.

## Scratch and Sniff

Italian scholar Francesco Petrarca (1304–1374) was in despair at the world he saw around him. Petrarch (as he is more commonly known) yearned instead for the glories of the ancients, and dreamed that they might somehow be recovered. Medievalist Dame Janet L. Nelson says that, according to Petrarch:

> the Middle Ages began when barbarians destroyed the Roman Empire c. 400, and the succeeding centuries of darkness (*tenebrae*) would last until western Europeans recovered the civilization of ancient Rome. . . . This fervent hope was what made Petrarch a historian—but a historian who passed over the medieval centuries in near-total silence: *"What else is all history if not the praise of Rome?"*[26]

Petrarch—the inventor of the term "Dark Ages"—was not alone in holding such a view. Nelson says Enlightenment hero David Hume (1711–1776) felt much the same, and that Edward Gibbon—perhaps the most famous historian ever to have lived—also saw the Roman Empire as humanity's high-water mark:

> After ending Volume III [of *Decline and Fall*] with the sack of Rome by the Goths in 410, Gibbon recorded "the darkest ages of the Latin world" . . . he had here to confront "the darkness of the middle ages" . . . "The sleep of a thousand years," thought Gibbon, could end only with a Petrarchian experience of rediscovery of Rome.[27]

This "sleep of a thousand years," then, is a bit of a recurring theme. We have now seen it in Draper and White, Freeman and Nixey, Petrarch and Hume—and, of course, the incomparable Gibbon. It was not, they make clear, a peaceful sleep.

Instead, it was riddled with nightmares.

Cities, even the richest and grandest of them that the Middle Ages could offer, were full of the depressed, dead, and dying. A recent BBC documentary, for instance, has presenter Dan Snow touring a reimagined fourteenth-century London.[28] In stark contrast to the glistening Gibbonian images of Rome, we are told that London boasted only seven—yes, seven—public toilets for the entire population; and that most of the waste simply ended up coating the already foul streets. The documentary, as it happens, is called *Filthy Cities*.

Lest anyone watching might fail to understand how horrible it all was, the BBC sent out some "scratch and sniff" cards to accompany the show. In one segment, Snow visits a modern sewage works, and viewers are invited to scrape away at the relevant box on their card. As they get a nice old whiff of human excrement, they are informed that this, for medieval Londoners, was effectively road-surfacing material. And that, it would appear, is what happens when religion takes over from science—people walk around, day to day, in their own bodily deposits.

In the monasteries, however, there was hope—for some still housed a handful of Greco–Roman texts. And yet, it was not to be. These precious books were erased, one by one, so that their pages could be reused—the newly blank leaves were then written over with Bible verses, or quotes from Church Fathers, or some other pious material.

One infamous incident of this—so infamous it has merited its own documentary[29] —was when an unknown thirteenth-century monk scratched away what we believe was the only remaining copy of a cutting-edge mathematical text by none other than Archimedes (c. 288–212 BC). The feckless friar then overwrote this potentially game-changing work with a prayer.

Such are the Middle Ages—the Dark Ages—of Draper and White. Here are some miscellaneous highlights (or, more accurately, lowlights) from *Conflict*:

> In the annals of Christianity the most ill-omened day is that in which she separated herself from science. . . . Universal history from the third to the sixteenth century shows with what result. The dark ages owe their darkness to this fatal policy.
>
> Personal cleanliness was utterly unknown; great officers of state, even dignitaries so high as the Archbishop of Canterbury, swarmed with vermin. . . . The streets had no sewers; they were without pavement or lamps. . . . How is it that the Church produced no geometer in her autocratic reign of twelve hundred years? . . . In Christian Europe there had not been

a cultivator of mechanical philosophy until Leonardo da Vinci, who was born A.D. 1452. . . . Christianity is responsible for the condition and progress of Europe from the fourth to the sixteenth century.[30]

And, as always, *Warfare* has *Conflict*'s back:

For twelve centuries, then, the physical sciences were thus discouraged or perverted by the dominant orthodoxy. . . . It came to be the accepted idea that, as soon as a man conceived a wish to study the works of God, his first step must be a league with the devil.[31]

When Carl Sagan wrote his multi-million-selling *Cosmos* in the 1980s, he included a history of science timeline for his readers—one which charted all of the great scientific developments made by humanity throughout recorded history. Between Hypatia in AD 415 and Leonardo Da Vinci in AD 1490, it goes rather quiet.

In fact, it is entirely blank.

Sagan explains, with obvious regret, that "The millennium gap in the middle of the diagram represents a poignant lost opportunity for the human species."[32]

Quite.

Still, we've been here before. So, when we discover that Michael H. Shank—emeritus professor of the history of science and co-editor of the multi-author, bang-up-to-date, 700-page *Cambridge History of Science: Middle Ages*—says "The crude concept of the Middle Ages as a millennium of stagnation brought on by Christianity has largely disappeared among scholars familiar with the period, but it remains vigorous among popularizers of the history of science"[33] we are not, perhaps, as surprised as we previously might have been.

## The Light Ages

Dr. Seb Falk has studied at Oxford University and is now researching and teaching at Cambridge University. His area of expertise? Medieval science.

Wait, hang on a second—medieval science? Surely the whole point is that there wasn't any, was there? After all, Sagan's timeline was empty. The Middle Ages were a time of superstition, of silly appeals to Scripture, of popes shooing away comets and praying away dragons, and of their faithful subjects spending most of their time covered in—well, let's be gentle about it—muck.

And yet, here is Falk, describing a medieval monastery:

> the cloister itself was decorated to reflect the breadth of monastic learning. . . . The windows included classical philosophers and poets, of course, but also medical thinkers, mathematicians like Pythagoras and Boethius. . . . Geometry and astronomy were represented by the totemic Greek masters Euclid and Ptolemy. . . . Significant recent thinkers in law and theology—Jewish as well as Christian theology—had their own windows, showing that the monks could appreciate both new ideas and the achievements of non-Christians.[34]

The name of Falk's book is telling: it is *The Light Ages*. In it, he chronicles how medieval universities taught the Greek and Roman classics alongside the latest astronomy and calculation; how medieval engineers built the first true and highly technical clocks; how complex numerical analysis was used to produce accurate calendars; how fresh water was piped into London gardens with springs as sources. His picture couldn't be more different from Sagan's. In fact, Falk speaks of finding "many handwritten mathematical tables" when researching scientific documents for his project. "No unicorns there," he says.[35] Indeed.

Falk is hardly out on a limb here. For, as it turns out, the long-standing myth of the Dark Ages was devastatingly debunked decades ago. But still it hangs around, like one of Dan Snow's bad smells.

So, let's try opening some windows—and see if we can't get rid of it.

## Read with a Critical Eye

We shall start by going back to the beginning—to the early Church's attitude toward Greco–Roman thought. The Draper–White line is that they were having none of it; that the likes of Paul, Tertullian, and Augustine had laid down the law when they pooh-poohed Plato and euthanized Euclid. But is that actually right?

Well, the Church, for a little while, tussled with two main questions: what should Christians be taught, and how should they think about what they were taught? This was a hugely important discussion; souls were potentially at stake. A handful of theologians, such as Tatian (AD 120–180), recommended avoiding everything pagan altogether—but these folk, in the end, proved to be the minority.

The majority position—and the one that quickly won out—was that learning the Classics was a good thing, provided it was done with care. In among the poems, the philosophy, the cosmologies, and the science, it was decided, were lessons that were of value, and lessons that were not. The Christian was to study it all, but also to weigh it for themselves—they were not to take it as a whole, unquestioningly, to heart.

Church historian Odd Magne Bakke has, rather helpfully for us, carefully reviewed what the various church leaders had to say on the matter. Here is his summary:

> The leading theologians in the Eastern tradition, from the beginning of the third century to the close of the fifth, all held that the Greek classics, which were the staple of the encyclical studies, contained elements valuable to Christians. . . . The same is true of Western fathers such as Jerome and Augustine . . . they unanimously emphasize the usefulness of this education. . . . They make the point that pagan literature should be read with a critical eye . . . the church did not found any alternative schools, either on the primary or the more advanced level.[36]

Note that even St. Augustine—the great villain of the Draper and White piece—makes Bakke's pro-pagan-philosophy list. Likewise, Tertullian— who is characterized as hating all things Greek—was nothing like as negative as many would have us believe. In fact, he writes: "Let us see, then, the necessity of literary erudition; let us reflect that partly it cannot be admitted, partly cannot be avoided. Learning literature is allowable for believers."[37]

And what of St. Paul? Well, he was himself a philosopher, as any trained reader of his work would immediately recognize. And, on the scientific front, he taught that study of the natural world could lead to a better understanding of God: "For since the creation of the world God's invisible qualities—his eternal power and divine nature—have been clearly seen, being understood from what has been made" (Rom. 1:20).

But if the Church was not vehemently opposed to Greek or Roman philosophy, and if it did not despise learning, then why did Christians burn libraries? Why did they murder mathematicians?

## An Alternative Saint

Perhaps, at this point, our level of mistrust of Draper and White has become so great that we might suspect Hypatia actually died peacefully in her sleep. Sadly, this is not the case. She was, as the various sources all state, violently killed at the hands of those who—rather unconvincingly—claimed to be disciples of the prince of peace.

Because of her death, Hypatia has risen up to become a transcendental figure; she has effectively been canonized as a sort of alternative saint—a martyr who represents reason standing firm against the dreaded arrows of blind faith. Intrinsic to Hypatia's God-versus-science iconography is her extraordinary prowess as a mathematician, and the burning of her scrolls, and the centuries of further cultural and intellectual devastation that followed her murder—and, of course, the profound sense of loss at what might have been had the Church never got hold of any power in the first place.

*Agora* and Nixey and the *Guardian* and Sagan offer modern retellings of the Hypatia story for a new generation—adding details to it as they do so. *Agora's* Hypatia put the Earth at the center of the solar system. Nixey's Hypatia ran a library with tens of thousands of scrolls. The *Guardian's* Hypatia was one of the top ten mathematicians of all time. Sagan's Hypatia was the last scientist before the Enlightenment.

And yet all this is precisely what it looks like—legend.

Firstly, and most importantly, Hypatia was not killed because she was rational, or because she was a scientist, or because she was anti-Christian, or because Christians were anti-science, or because she was an atheist, or because she was a woman. She was a neo-Platonist who believed in the supernatural, and she was killed because she inadvertently got caught up in a complex political power struggle in a violent city famed for its rioting and brutality, and in which murder was the order of the day.[38] Here is classicist John Dickson:

> The reason the murder of Hypatia simply can NOT be about the church's opposition to secular learning—to "science"—is at least threefold: (a) our best contemporary source—Socrates Scholasticus—is a Christian, and yet he praises her as the best philosopher of the day and deserving of fame, (b) some of her admiring students were Christians (e.g. Synesius and his brother Eutropius, both of whom became bishops), and (c) scholarly

inquiry continued in Alexandria long after Hypatia, including that of the devoutly Christian John Philoponus.[39]

Investigate yet further, and the legend gets even more badly exposed. The mathematician Michael Deakin wrote a careful biography of Hypatia, and found that the vast majority of the material he came across about her is "fanciful, tendentious, unreferenced or plain wrong." He continues:

> What we know of Hypatia is little enough; what we know of her Mathematics is only a small subset of that little. . . . We have no evidence of research Mathematics on the part of either father or daughter. What we can reconstruct of their Mathematics suggests to us that they edited, preserved, taught from and supplied minor addenda to the works of others. A great deal of Theon's work survives and at most a small part of Hypatia's. In other words Theon was seen as the better text-writer.[40]

The *Guardian*'s placing of Hypatia on its top ten list is unjustifiable on her mathematics alone, for we have next to nothing to judge her on—it seems her murder has given her mathematical reputation a significant boost.

*Agora*, for what it is worth, simply made up the bit about Hypatia figuring out the solar system—she did nothing of the kind, and those discoveries weren't made until the sixteenth and seventeenth centuries. Still, it is a movie, so artistic license is permitted. Nixey's claim about the vast number of books burned in the *Serapeum*, however, is supposed to be a factual one. Does it stand up to further scrutiny? Well, philosopher David Bentley Hart thinks not:

> As it happens, we have fairly good accounts of that day, Christian and pagan, and absolutely none of them so much as hints at the destruction of any large collection of books. Not even Eunapius of Sardis—a pagan scholar who despised Christians and who would have wept over the loss of precious texts—suggests such a thing. This is not surprising, since there were probably no books there to be destroyed. The pagan historian Ammianus Marcellinus, describing the Serapeum not long before its demolition, had clearly spoken of its libraries as something no longer in existence.[41]

Hart, in fact, reckons he can pinpoint the source of this myth:

The truth of the matter is that the entire legend was the product of the imagination of Edward Gibbon, who bizarrely misread a single sentence from the Christian historian Orosius, and from it spun out a story that appears nowhere in the entire corpus of ancient historical sources.[42]

Nixey and Hart, then, differ on the contents of the *Serapeum*. But why? Either there was a library, or there wasn't. Either it did have books in it, or it didn't. This is not a matter of opinion. It is a matter of historical fact.

So how can there be disagreement?

## Three Types of Dream

Roger Bagnall is a very well-established and highly decorated professor of ancient history—and his specialism is long-lost collections of books. One of the things he has learned along the way is that sorting historical truth from historical fiction is not always as easy as those outside of the discipline might have thought.

And he has a great example to prove it.

Centuries before the pseudo-events portrayed in *Agora*, there was already a library in Alexandria—perhaps the most famous library to have ever existed. All sorts of assertions are made about it: that it held every book written up to that point; that it was the center of all the world's wisdom; even that, if it had not been destroyed, we might have got to the moon before the turn of the first millennium AD.[43]

Here, for instance, is Sagan's take on it in *Cosmos*:

Only once before in our history was there the promise of a brilliant scientific civilization. . . . it had its citadel at the Library of Alexandria, where 2,000 years ago the best minds of antiquity established the foundations for the systematic study of mathematics, physics, biology, astronomy, literature, geography and medicine. . . . From the time of its creation in the third century B.C. until its destruction seven centuries later, it was the brain and heart of the ancient world.[44]

Bagnall, however, suggests that people calm down somewhat—for, as we saw with Hypatia, we don't know quite as much about all this as many seem to think we do:

No one, least of all modern scholars, has been able to accept our lack of knowledge about a phenomenon that embodies so many human aspirations. In consequence, a whole literature of wishful thinking has grown up, in which scholars—even, I fear, the most rigorous—have cast aside the time-tested methods that normally constrain credulity, in order to be able to avoid confessing defeat. . . . I shall talk about three types of dreams that have beguiled commentators ancient and modern: dreams about the size of the Bibliotheca Alexandrina; dreams about placing the blame for its destruction; and dreams about the consequences of its loss.[45]

Bagnall goes on to explain that we don't even know how or when or by whom the original library was built, let alone which documents were in it. Oft-repeated claims that there were 700,000 or 500,000 or even 200,000 scrolls are entirely beyond the pale, he says: such "outlandish" figures "do not deserve any credence" and "lead to impossibilities and absurdities." He puts the total number as likely nearer to ten or fifteen thousand—and even this is assuming multiple copies of most works.

What's more, despite Sagan appearing to think that the original *Bibliotheca Alexandrina* and Hypatia's *Serapeum* were one and the same (hence his "seven centuries"), Bagnall makes it clear that they were two entirely separate institutions—the former a long-distant memory by the time Theon and his daughter were around. The *Serapeum* perhaps had a handful of scrolls in it for classes—but it may well have housed none at all. After all, there are no ancient records, anywhere, of it holding books or of those books being burned—not by Christians, and not by anyone else.

It is beginning to look like the "tens of thousands" of manuscripts Nixey believes in are just as much an article of the conflict thesis faith as Hypatia herself has become. Bagnall again:

Passions still run high on this matter . . . I wrote an article on the Alexandrian Library. . . . The editor did not like my caution about the accounts of the destruction of the Library and, without telling me, rewrote the article to blame everything squarely on the Christians. Whether he hated Christianity or just liked a simple story line, I do not know.[46]

Still, there is one thing we can be certain of—those libraries aren't there now. They weren't there at the end of the fifth century either. What's more, any scrolls in use at Hypatia's time were highly susceptible to climatic wear and tear, and would have rotted away—even in storage—long before AD 600.[47]

Remarkably, though, we know what was written in some of them. How?

## Grayling Versus Holland

Anthony Clifford Grayling is a stalwart of the British intelligentsia. A philosopher and a prolific author, he has written more than thirty books and is currently master of London's New College of the Humanities. In his sweeping 2019 volume, *The History of Philosophy*, he appears to line himself up somewhat with Draper and White's version of events. Grayling says that, from Augustine onward:

> A vast amount of the literature and material culture of antiquity was lost, a great deal of it purposefully destroyed. Christian zealots smashed statues and temples, defaced paintings and burned "pagan" books. . . . It is hard to comprehend, still less to forgive, the immense loss of literature, philosophy, history and general culture this represented.[48]

This assertion made classicist Tom Holland rather angry. We know this because, at the time, he was sat less than four feet away from Grayling, and the two men were being filmed. Holland's response was direct:

> Anthony is a great scholar and professor, and you would think that this would be a simple thing for him to go and check. This is a myth that essentially is propagated in the eighteenth century. The figure who underlies it is Gibbon.[49]

The truth, he said, is not only that Christians didn't burn pagan books—it is that they were the ones who kept them alive:

> Monks were systematically copying, they were copying Virgil in the Latin West, they were copying Horace, they were copying Ovid, that's why they survived. . . . In Constantinople they were copying Homer, they were copying Herodotus, they were copying Thucydides, that's why they survived.[50]

Indeed, the major point missed by those complaining about twelfth-century monks erasing Archimedes is this: how did twelfth-century monks have Archimedes—which, by then, would be 1400 years old—in the first place? The answer is a simple one: their predecessors had been carefully studying and producing copy after copy of it, generation after generation—hardly the behavior of an unscientific bunch.

Perhaps we can let Holland sum up:

> The idea that there was a systematic campaign by evil Christians to eliminate the legacy of classical civilisation could not be less true—and this is so clear and transparent a historical fact that it stupefies me that Anthony could even begin to think otherwise.[51]

Well, then. That appears to be that.

## God of the Gap

So, if Christians didn't do what Draper and White would have us believe—if they didn't burn up the Greco-Roman intellectual world; if they thought Greco-Roman education was useful; if they played the major role in keeping it available; if they decorated their monasteries with famed mathematicians and scientists; if they were open to ideas from other cultures—then why was there a millennium of darkness between AD 500 and AD 1500? What happened in between Rome and the Renaissance? Why does Sagan's empty timeline just sit there, all white space?

Well, in the last half-century or so (at least), scholars have looked again at this gap—and they have been rather busy filling it in. We already had an example of this in Falk's *Light Ages*, with its clocks, mathematical tables, and calendars—and these turn out to be just the tip of an intellectual iceberg.

In *God's Philosophers*, James Hannam tells the stories of many major medieval thinkers and their contributions to human progress. He relates how Christians committed themselves to the study of any Greek science or philosophy they could get their hands on, and that they energetically pursued studies such as astronomy, mathematics, and medicine. In his introduction, we come across some rather familiar names:

> Daniel Boorstin's history of science *The Discoverers* referred to the Middle Ages as "the great interruption" to mankind's progress . . . Charles Freeman wrote . . . that this was a period of "intellectual stagnation" . . . John William Draper and Thomas Huxley introduced this thesis to English readers in the

nineteenth century. It was given intellectual respectability through the support of Andrew Dickson White. . . . But anyone who checks his references
will wonder how he could have maintained his opinions if he had read as
much as he claimed to have done.[52]

In the Draper–White–Gibbon–Petrarch–Nixey–Freeman–Boorstin–
Sagan–*Agora*–Grayling gap, as one might call it, researchers keep on finding
people who are not supposed to be there: people who resolutely believe in
God, live squarely in the Dark Ages, and are doing some pretty clever open-
minded thinking.

Boethius (c. 477–524) and John Philoponus (c. 490–570), for example,
analyzed and developed Greek philosophy and Greek physics, respectively.
Philoponus discussed the idea of forces being the true cause of motion and
even argued, against what he had read in Aristotle, that different weights
would fall at the same speed.[53]

The Venerable Bede (c. 672–735) was a historian and natural philosopher whom Allan Chapman refers to as "Britain's first astronomer of
international standing."[54] Bede did not confine himself to matters of cosmology: when wondering how the sea could be salty when it was fed by
rivers, he dismissed the old Roman theory he found in Pliny the Elder (AD
23–79). The classical case was that there were tunnels under the sea by
which the fresh water returned to the rivers. Bede, employing empirical
evidence, disagreed: "But fresh waters flow above salt waters, for they are
lighter; the latter certainly, being of a heavier nature, better sustain the waters poured over them."[55]

Alcuin of York (c. 735–804) was schooled by one of Bede's students, and he
delighted in education of all kind. He is famous, alongside his more scholarly
and clerical work, for inventing logic puzzles specifically designed to train
young minds in reason—such as this one:

> A man had to take a wolf, a goat and a bunch of cabbages across a river. The
> only boat he could find could only take two of them at a time. He had been
> ordered to transfer all of these to the other side in condition. How could
> this be done?[56]

Some of the other names involved in medieval science include: Leo the
Mathematician (c. 790–869), an archbishop of Thessalonica who wrote an
encyclopedia of medicine; Gerbert of Aurillac (946–1003), a mathematician

and astronomer who was eventually made pope; Constantine the African (c. 1020–1087), a physician who translated Hippocrates and Galen; Adelard of Bath (c. 1080–1152), who penned *Questions on Natural Science*; Robert Grosseteste (c. 1168–1253), bishop of Lincoln and hands-on experimentalist who uncovered many of the properties of light; St. Albert the Great (c. 1200–1280), an expert in logic, psychology, metaphysics, meteorology, mineralogy, and zoology; and Roger Bacon (c. 1214–1294), perhaps the first truly mathematical physicist and a serious proposer of hot air balloons, motorboats, and flying machines.

Then, from AD 1300 to 1500, there were such luminaries as William of Ockham (c. 1287–1347), a scientist and philosopher who is quoted in theoretical physics books to this day; Jean Buridan (c. 1301–1358), whose logic regularly features in modern economics and psychology; Thomas Bradwardine (c. 1300–1349), who calculated the mathematics of acceleration; Nicole Oresme (c. 1320–1382), who represented motion graphically; and Nicholas of Cusa (1401–1464), who first took people's pulse rates, and who theorized the motion of the Earth through space.

These lists are far from being exhaustive, of course. Each name here (and many others besides) comfortably merits a book of their own to detail their extensive work—indeed, most of them now have more than one. Almost every day new discoveries are made about the science going on in the gap.

Progress was not limited to scientific thinking, either—there was plenty of scientific doing, too. The specialist in medieval technology, George Ovitt, picks out multiple life-changing devices and techniques such as the magnetic compass, the stirrup, crop rotation, water- and wind-powered mills, spectacles, flying buttresses, ploughs, horse collars, the crossbow, batch production of wool, cannon, and the new skeleton-style construction of stronger and lighter ships.[57]

What's more, Ovitt suggests that these tinkerers of Christendom "were moved to invention out of some restless spirit of creativity," whereas their Roman forebears—the ones so vaunted by Gibbon et al.—"displayed little interest in original invention."[58]

So the gap, then, is pretty full—full of God-fearing and thoroughly medieval minds that were making scientific, philosophical, and technological progress right across the board. Sagan missed several tricks, it would seem.

But what about the death, and the filth, and the superstition that dominated the Dark Ages? What about the monster-killing, comet-dismissing popes? What about the unicorns?

## goop

Well, there is no denying it: superstition in the Middle Ages was rife. Peter Dendle, an expert on folklore, lays out just a few of the many examples:

[There were] unlucky "Egyptian" days, and anyone daring to have his blood let or to eat goose meat on those days would, according to some sources, die shortly thereafter. . . . Springs, brooks, rivers, and wells were thought to be inhabited by sprites or spirits. . . . People were sometimes believed to turn into wolves during a full moon. . . . one can easily generate an amusing list of quaint beliefs and absurd practices.[59]

But matters are not quite so simple, and Dendle cautions us against making snap judgments. Greek and Roman giants such as Plato, Theophrastus, Cicero, and Plutarch, he says, also describe a world full of ungrounded fantasies. They tell of astrology and divination, of panic caused by sweating statues, and of the simultaneous defecation of two oxen being a dangerous omen.[60]

Superstition, then, pre-dates the Dark Ages. And, more to the point, it is still very much with us now—right here in the supposedly rational and post-Enlightenment world of modernity.

Take goop, for instance. Marketed as a health and wellness company fronted by Oscar winner Gwyneth Paltrow, goop had to pay out a large sum of money in September 2018 after a ruling it had made "unscientific claims." These included selling a flower essence which could apparently "cure" depression, and eggs cut from jade which could—when placed in the right spot anatomically—help balance a suffering woman's hormones.[61]

Undeterred by this lawsuit, the brand pressed on—and it now has a Netflix hit with *the goop lab*. A documentary series about goop's "science," it features one episode in which a certain Dr. John Amaral treats people by manipulating their "energy fields." He does this using only his bare hands, and can do so from as far as 6 feet away: "I have a hypothesis: If you just change the frequency of vibration of the body itself, it changes the way the cells regrow."[62]

Talking a thrilled Paltrow through it all, Amaral cites famous experiments from "quantum physics." Well, Professor Phil Moriarty of Nottingham University just happens to be a quantum physicist—and, somewhat wonderfully, his live reaction to this particular episode of *the goop lab* is available for all to watch on YouTube. It contains such highlights as "this is bonkers," "I've only got to the titles and I'm already p***** off," and "this is just nonsense, this really is just nonsense."[63]

For what it's worth, goop has half a million followers on Facebook. The company is valued at a quarter of a billion dollars.[64] Here is Dendle once more: "It is not a given that, even from an absolute standpoint, people in the Middle Ages entertained more superstitions than, for instance, twenty-first century Americans do."[65]

Yes, the Middle Ages were superstitious. So were the Greeks and Romans. So are people now.

## Popes and Unicorns, Revisited

Just as superstition seems to be a human constant, so do stories of mysterious beasts. White made it sound like only medieval folk were enchanted by such ideas, but that is far from the truth. The ancient Greeks went in for them big time, with Hesiod (fl. c. 700) writing about dog-headed tribes and Empedocles (c. 494–433 BC) making seemingly serious claims about disembodied hands and feet bounding around of their own free accord. Even Herodotus (c. 484–425 BC) leaves room for the existence of half-lion-half-eagle griffins—although, to his great credit, he appears to favor the side of skepticism.[66]

And, of course, we ourselves indulge in such fantasies—rarely an evening goes by without a mainstream TV channel somewhere searching for the Loch Ness monster, or the Abominable Snowman, or extraterrestrials who, having built the pyramids, are occasionally popping back to the Earth to abduct and probe yokels. It would appear that we still, in our own distinct way, very much believe in unicorns.

Well, OK, but what of the unicorns in the Bible? Wasn't White's point that otherwise sensible people like Isidore were being misled by Scripture itself?

An unlikely candidate to sort this particular chestnut out is arch-atheist and science writer Isaac Asimov. In his *Guide to the Bible*, Asimov soundly debunks the idea that the biblical writers ever claimed such a creature existed—and he also shows that they never sought to convince anyone else that it did, either:

The Hebrew word represented in the King James Version by "unicorn" is re' em, which undoubtedly refers to the wild ox (urus or aurochs). . . . When the first Greek translation of the Bible was prepared about 250 b.c. the animal was already rare in the long-settled areas of the Near East and the Greeks, who had had no direct experience with it, had no word for it. They used a translation of "one-horn" instead and it became monokeros. In Latin and in English it became the Latin word for "one-horn"; that is, "unicorn."[67]

The mystery of the Scriptural unicorn, then, is one of a bad translation. When the Bible spoke of a "unicorn"—which it only ever did in a few translations—it was speaking of a wild ox. And it never mentioned somersaults. Sorry, Cosmas—wrong again.

Isidore and other readers then presumably conflated this animal with the older idea, present in Greek texts, of a single-horned horse that lived in India. They were hardly being naïve—for on what grounds should they not believe in such an easily imagined creature? After all, there are far, far stranger creatures out there that really do exist—the Alaskan wood frog, for instance, freezes solid all winter long and thaws out again in the spring. And, if folk in the Middle Ages wanted to fantasize further about unicorns—which they did, a great deal—is that really any different to our documentaries on Nessie?

That's unicorns dealt with, then. What about popes?

Well, the "pope excommunicates comet" storyline from *Conflict* also turns out to be rather disappointing in reality. In that it didn't happen. There is no mention whatsoever of the idea in any primary source at all and the whole thing, upon further investigation, disappears in a puff of smoke as a nineteenth-century urban myth. In fact, the tale's lack of provenance was exposed by careful historians within Andrew Dickson White's own lifetime—and not just once, but twice.[68]

And, while we are still pontiff-icating, there is also (sadly) no primary evidence that a pope ever killed a laser-eyed basilisk in Rome—or anywhere else, for that matter. These travelers' tales were written as entertainment for the masses, not careful history for the academy—and their readers treated them as such. Eugene Roger's journal is to medieval science what *the goop lab* is to quantum physics—in that it really doesn't tell us anything about it.

Which leads us, in fact, to an interesting question: were there any Professor Moriartys around in the Middle Ages? Were there people prepared to challenge irrationality or silliness when they saw it? The answer is yes—there were.

## Medieval Moriartys

In Hannam's gap-stuffing *God's Philosophers*—which was, incidentally, nominated for the Royal Society's Prize for Science Books—he discusses the mighty theologian Thomas Aquinas's (1225–1274) thoughts on the practice of astrology:

> [Aquinas's] views represent something approaching the medieval consensus and are worth quoting: "If anyone attempts from the stars to foretell future contingent or chance events, or to know with certitude future activities of men, he is acting under a false and groundless presumption, and opening himself to the intrusion of diabolic powers. Consequently, this kind of fortune telling is superstitious and wrong. But if someone uses astronomic observation to forecast future events which are actually determined by physical laws, for instance drought and rainfall, and so forth, then this is neither superstitious nor sinful."[69]

Aquinas, it needs to be pointed out, was massively influential—indeed, he is perhaps the most influential thinker in the entire history of the Church, with the possible exception of Augustine. This matters because, as Hannam records, his views were much more Moriarty than goop:

> In particular, Thomas stood up for the doctrine of secondary causes as a valid way for a Christian to investigate the world. He did not accept that it was impious to say that a plague was caused by a disease rather than attributing it directly to the will of God.[70]

Indeed, the average sick person throughout the Middle Ages would both pray *and* seek medical treatment from wherever they could get it. In his 2019 book *The Middle Ages: Facts and Fictions*, medievalist Winston Black says:

> We first need to dismiss those mythmakers who claim that medieval people resorted only to prayer and magic to treat disease. Medieval medicine recognized most diseases as having natural causes, treatable by natural remedies, which could be understood and prepared by men or women, the learned or the uneducated.[71]

The pray-and-get-treated approach is consistent with what St. Paul suggested in Scripture (as we saw in our previous chapter) and is what remains, by far, the most common practice of Christians today. The assertions coming from Draper and White that prayer somehow replaced medicine entirely are, as Black makes abundantly clear, wholly untrue.

Another medieval Moriarty was St. Bernard of Clairvaux (1090–1153). This French abbot takes a real beating from White in *Warfare* as being a typically dogmatic Christian and, correspondingly, a curse to rationality. And yet, in Bernard's *Apologia* of AD 1125, there is a passage about the beasts of the bestiaries so scathing that White could probably have written it himself:

> Here [in the cloisters] on a quadruped we see the tail of a serpent. Over there on a fish we see the head of a quadruped. There we find a beast that is horse up front and goat behind, here another that is horned animal in front and horse behind. . . . Good Lord! If we aren't embarrassed by the silliness of it all, shouldn't we at least be disgusted by the expense?[72]

Not everyone in the Middle Ages, then, was a superstitious nutcase—at least, there were no more of them then than there are now. And, more often than not, those questioning the more eccentric ideas of the time were the supposed anti-science villains of *Conflict* and *Warfare*—the dedicated clergy.

## Sweat, Urine, Feces, and Decay

So is nothing about the Dark Ages myth sacred? Surely, if we know anything at all, we know that the medieval period stunk, and stunk pretty bad—didn't it? After all, those scratch and sniff cards provided by Dan Snow didn't bother with patches which smelled of antiseptic. Or of toothpaste. Or of deodorant. Or of soap.

Even this core belief, though—a key line in the Gibbonian creed—requires some re-evaluation. Firstly, things weren't all that olfactorily pleasant in the glorious ancient world. Here's Odd Bakke again:

> People in antiquity were dependent on chamber pots and on holes in hills that functioned as latrines; the contents had to be emptied into open sewers, into which other domestic rubbish also was thrown. In many cases, however, people could not be bothered to do this, and simply emptied their chamber pots out of the windows during the night. Ventilation in the apartments was ineffective. . . . As Rodney Stark says, "The smell of sweat,

urine, feces, and decay permeated everything." He points out that things were not much better out of doors on the street: "Mud, open sewers, manure, and crowds. In fact, human corpses—adult as well as infant—were sometimes just pushed into the street and abandoned."[73]

Medieval London, by comparison, doesn't come off quite as badly as we might have thought—in fact, in many ways, it was a significant improvement. Norwegian historian Dolly Jørgenson explains that there were strict laws about what you could do with your mess; that these were enforced, when necessary, by the courts; that breaches of protocol were rare; and that specific taxes were collected to keep the river and streets clean. She rails against common and lazy claims that "hygienic conditions fell far below the standard of Imperial Rome,"[74] and seeks to set the record straight:

> Certainly there were transgressions of waste disposal norms in the Middle Ages, but just as we have people who litter or throw a sack of garbage on a countryside lane today, those were the exception, not the rule. Medieval city dwellers did not trample through ankle-deep refuse in the street every day—they would have found that as loathsome a prospect as Snow did in *Filthy Cities*.[75]

All things considered, it seems that the Middle Ages—and the often clever, resourceful, and sensibly pious people who lived during them—deserve a genuine rethink on the popular level. Perhaps, instead of our children being told some rehashed Petrarchan tale about how everyone was stupid, horrible, and smelly, they could be taught about some of the heroes who threw themselves into science, philosophy, and math. Perhaps they could be treated to live demonstrations of ingenious medieval inventions—and asked if they could ever have come up with the same ideas.

Perhaps, even, they could be set some of Alcuin's fun, imaginative, and more-than-a-millennium-old puzzles—for it might help to sharpen up their twenty-first-century brains.

## The Writings of Augustine

One last task remains before us in this chapter: the gentle rehabilitation of a battered saint.

Augustine was blamed by Draper for the Dark Ages; Augustine was mocked by White for his credulity. The picture the pair paint is of a dangerously useless but influential man with his eyes stuck in Scripture and his ears deaf to reason. But they are wrong—and it is really rather easy to prove it.

Draper, for instance, had said that any medieval astronomical question "was at once settled by a reference to the writings of Augustine," and that the Middle Ages had therefore "not produced a single astronomer."

This, however, is bunk—on both counts. Firstly, there were thousands of astronomers throughout the length and breadth of Christendom, many of whom were monks, or cardinals, or even popes. Some made their own machines called astrolabes to track the planets and stars; others wrote manuals on how to build them; still others wrote instructions on their correct use—all this is detailed, complete with images of manuscripts, in Falk's *Light Ages*. Secondly, Draper was wrong to imply that Augustine was hopelessly unscientific—for he was actually rather capable in that regard.

Take the widespread and ancient practice of astrology, for instance. Augustine, in his mammoth theology *The City of God*, takes it to task—and he uses the power of reason to do so.

Twins, Augustine says, are born under the same stars, and yet their lives can dramatically diverge—an awkward fact for the astrologer, and one which our theologian thinks is already enough to entirely undermine the practice. What's more, he references both Cicero and Hippocrates on the matter— giving a lie to any idea that he ignored the classics.

Hippocrates, according to Cicero, had famously decided that two brothers must be twins when they fell ill at the same time—but Augustine, mentioning this, argues that one event does not necessarily follow from the other. Instead, he thinks the better explanation is that both boys lived in the same environment, ate the same food, and were treated alike by their parents. The bishop even goes on to say that he thinks Hippocrates, if the Greek physician were around to hear it, would agree with the logic of his newer assessment.[76]

This is all well and good, but what of White's accusations about immortal peacocks and transmogrified donkeys? Is there anything that can be said in the Church Father's defense about these bizarrest of bizarre charges?

Let's deal with the cheese-donkeys (mascarponies?) first. Augustine, again in *City of God*, says this story had indeed been passed on to him—that cheese had turned people into asses somewhere in Italy. His response is as follows:

> These things are either false, or so extraordinary as to be with good reason disbelieved. . . . I cannot therefore believe that even the body, much less the mind, can really be changed into bestial forms and lineaments by any reason, art, or power of the demons.[77]

So, in short, he rejected the report. However, he does not go on to rule it out entirely—for he believes that God can do anything. This approach—one of rational skepticism paired with an open mind—would go on to become hugely important in the unfolding story of science, as we shall see later in this book.

And now peacocks. The legend that the bird's meat did not go moldy was hardly a new one—it had circulated among the ancient Greeks. Augustine, however, doesn't just take their word for it:

> This property, when I first heard of it, seemed to me incredible; but it happened at Carthage that a bird of this kind was cooked and served up to me, and, taking a suitable slice of flesh from its breast, I ordered it to be kept, and when it had been kept as many days as make any other flesh stinking, it was produced and set before me, and emitted no offensive smell. And after it had been laid by for thirty days and more, it was still in the same state; and a year after, the same still, except that it was a little more shriveled, and drier.[78]

Any biologist, chemist, or physicist worth their salt will immediately recognize the passage above for precisely what it is: a fully fledged—and surprisingly modern—science experiment. Many popular histories of science suggest that experiments of this form only first appeared in the 1600s with the likes of Galileo, and Boyle, and Hooke. Yet here we are, at around AD 420, and Augustine has beaten the lot of them to it. He has a hypothesis, he has a method, he has data, he has a conclusion. By referencing other meats, he has even built in a control variable of sorts.

What we have got here, effectively, is a dogmatic theologian using the scientific method more than a thousand years before the so-called scientific revolution. Draper and White were horribly mistaken. Augustine's writings didn't lower the bar, they raised it—and, in doing so, set high standards for the non-gap that was to come.

## A Poignant Lost Opportunity

Frustratingly, despite all the contrary evidence on offer, the miserable middle ages picture just won't go away. Nixey's book, for example, was widely panned by classicists from universities all over the world, but was lavishly praised in high-circulation newspaper reviews. *Agora*—which has Oscar-winning superstar Rachel Weisz playing Hypatia—is always going to hit a far bigger audience than Bagnall's carefully corrective article in the *Proceedings of the American Philosophical Society*.

Amenábar's inspiration for *Agora* was the TV version of Carl Sagan's *Cosmos*—which was itself watched by half a billion people, exposing each and every one of them to the gap.[79, 80] Just like so many of Draper and White's other ideas, then, it is now deeply embedded in the minds of the public.

All of which is incredibly frustrating to those academics who know what they are talking about. Here is a telling passage from the comprehensive *Cambridge History of Science*: "The timeline reflected not the state of knowledge in 1980 but Sagan's own 'poignant lost opportunity' to consult the library of Cornell University, where he taught."[81]

There is some sort of strange irony in the fact that Sagan worked at Cornell, the very same university founded by Andrew Dickson White. There is another in the fact that the libraries there—one of which still bears White's name—could actually, if they had been checked, have put an end to the gap once and for all.

Had Sagan done just a little more research, he could have told 500 million people that there actually was plenty of science and technology in the Middle Ages. He could have told them that medieval folk were often quite clever, and inventive, and practical. He could have righted Draper and White's wrongs.

In 2014, however, Sagan's *Cosmos* franchise was given a chance to set the record straight: America's PBS commissioned a TV series of the same name, intending it to be a part-sequel-part-reimagining of the original for a new generation.

The resulting show—*Cosmos: A Spacetime Odyssey*—went down a storm with both fans and critics alike, and won four Emmys, including one for its bold new script. Does this mean the revised and updated *Cosmos* had learned the lessons of its predecessor? Had it finally moved on from Petrarch and Gibbon, and from Draper and White?

We will answer these questions in the next chapter. For now, though, here is a clue: the presenter this time around was Neil DeGrasse Tyson, whom we previously encountered in Chapter 3. Back then we found him confidently informing his huge Twitter audience that, although the ancient Greeks had figured out we lived on a globe, such knowledge had been "lost to the Dark Ages."

It doesn't bode well, does it?

John William Draper (1811–1882) and Andrew Dickson White (1832–1918) wrote the bestselling *History of the Conflict Between Religion and Science* (1874) and *A History of the Warfare of Science with Theology in Christendom* (1896), respectively. Since then, they have been known in the scholarly literature as co-founders of the conflict thesis: the notion that science and religion are fundamentally and irrevocably at war.

The French Revolution sought to replace traditional Catholicism with new "religions"—the Cult of Reason and the Cult of the Supreme Being. The cathedral of Notre Dame in Paris was rededicated in 1793 to the Cult of Reason, and the goddess of liberty was worshipped where the altar had once stood.

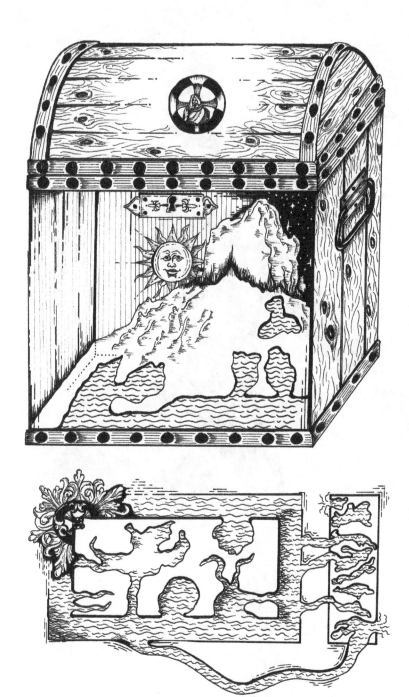

Cosmas, a sixth-century convert to Christianity, tried to use Scripture to devise a correct picture of the Earth. His work was either ignored or ridiculed by his peers and had no impact on Christianity as a whole—despite claims to the contrary from both Draper and White.

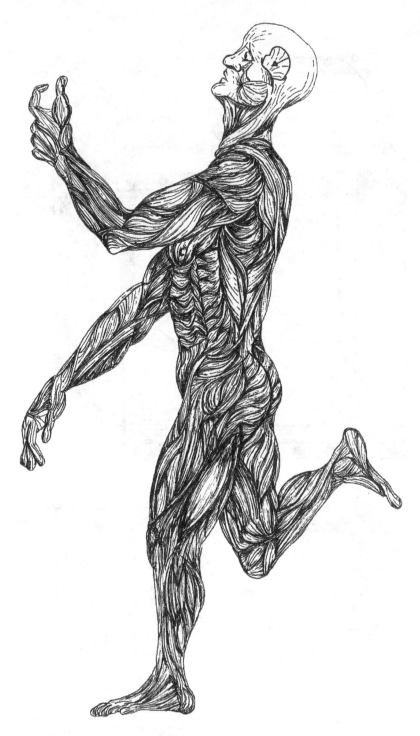

Andreas Vesalius (1514–1564) expertly dissected the human body, taking great care to document its inner workings. His work contained detailed woodcuts by a group of skilled artists and advanced anatomy significantly. Such study was legalized, enabled, and supported by the Church from the twelfth century onward—overturning the bans on the practice which had originated in the ancient world.

The popular image of the Middle Ages as the "Dark Ages"—a miserable and fearful period dominated by death, disease, irrationality, fecklessness, and stupidity—is far from accurate. In reality, much scientific thinking was present; and advances in technology, logic, and philosophy were quite considerable. Much of this was deeply integrated with the Church.

French astronomer Camille Flammarion (1842–1925) included a novel engraving in his 1888 book *L'atmosphère: Météorologie populaire*, reminding his readers that previous generations had imagined the universe as a series of solid spheres. Thankfully, he implied, science had risen up to give us a new and reliable way of chasing down the truth—and the image soon became linked to myths about Copernicus, Bruno, Galileo, and more.

Far from being a hard-nosed atheist, Draper considered himself to be an enlightened Christian, following the faith as it had been in its original and pure form. When he began experimenting with photography, one of the very first images he took was of Raphael's *The Deposition of Christ* (1507), in which Jesus is being mourned soon after his death.

Many of the key thinkers in the seventeenth-century world of science were heavily influenced by the biblical story of the Fall. The likes of Francis Bacon (1561–1626) and Robert Hooke (1635–1703) wrote that science was a gift from God to help humankind reverse the devastating effects of Adam's sin. They also believed that the fallen human mind would need support from experimental work since it was unreliable on its own. As a result, this Christian dogma gave great energy to what we now call the "scientific method."

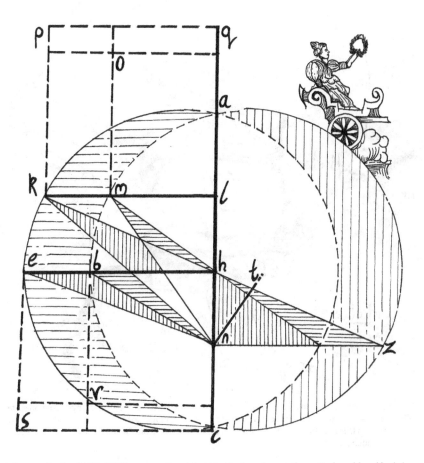

Johannes Kepler (1571–1630) was convinced that God had made a mathematical world and had also gifted his people with mathematical minds. He believed that by investigating the heavens he might learn more of God and worship him more fully. The outcome of such an approach was his famed "three laws" of planetary motion—which, in turn, enabled the similarly minded Isaac Newton (1643–1727) to formulate the law of gravity. The picture above shows Kepler's orbit of Mars.

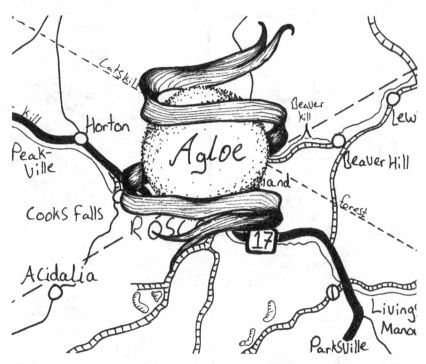

A fictitious town invented to catch copyright frauds back in the 1930s, Agloe has since taken on a life of its own. Like the conflict thesis, it was birthed from the imaginations of its creators and went on to fool many into thinking that it was real. In 2014, it was removed from Google Maps. Will the conflict thesis last much longer?

# 6

# Myths, Myths, Everywhere

## Feelings of Loneliness and Emptiness

"As feelings of loneliness and emptiness slowly begin to surge upon me," wrote Yusaku Maezawa, "there's one thing I think about: Continuing to love one woman."[1]

Maezawa had recently broken up with his girlfriend. He was heartbroken. In desperation, he threw out a worldwide appeal for a special someone to come and heal his hurt: "I want to find a life partner," he implored womankind everywhere. "Come to the moon with me?"[2]

This was no poetic exaggeration on the Japanese businessman's part; he had the flight booked already. Thanks to fellow billionaire Elon Musk's SpaceX program, Maezawa and his lucky girl would fly around the moon in 2023. What the lovelorn adventurer had paid for this twin ticket was not disclosed—but Musk, one of the world's three richest entrepreneurs, called it "a lot of money."[3]

Yet why, if money is no object, stop at the moon? Why not aim further? Well, those are precisely the thoughts of Musk himself, and of his assorted competitors: the race to Mars is on.

Right now, various governmental agencies are either teaming up with or working against private companies to be the first to put people on the red planet. A one-way trip is likely to take the better part of a year, even though Mars, cosmologically speaking, is pretty close to us. A journey to the sun would take only a little longer—but would, of course, burn up the unfortunate tourists (even if they did go at night).

The far reaches of our one-sun-and-eight-planets solar system are also just about within reach, although the impracticalities of a decade-plus of travel would have to be overcome. It is only when we begin to think about getting to the next star system along, Alpha Centauri, that a real sense of universal size begins to dawn on us. For to travel there—to our nearest stellar neighbors— would take a thousand centuries.

Yet even that impossibly huge star trek is just "round the corner" as far as space is concerned. Our sun is one of more than one hundred thousand million stars in our galaxy alone. And the universe—or, at least, the bit of it that we can see—holds more than a hundred billion galaxies just like ours. This makes for a discombobulating estimate of there being somewhere around 100,000,000,000,000,000,000,000 stars out there, and very possibly orders of magnitude more—many of which boast planets of their own. And, if we ask how much further space on goes beyond what we can see, the answer is a simple one: we don't know.

All this is rather different from ancient Greek naturalist Aristotle's picture of the cosmos. In his version, the Earth sat in the center with the sun and planets orbiting around it. Then, past those, lay the rotating sphere of the fixed stars. Pretty much everyone was happy with this for pretty much 2000 years, and most of the astronomical predictions that came from it were useful enough. So, if it ain't broke, why fix it?

But, of course, it *was* broke.

## First of Those Who Should Be Mentioned

The unexpected discovery of the true vastness of space is the subject matter in episode one of the Tyson-fronted reboot of Sagan's *Cosmos*. The writers of this 2014 series had quite a task on their hands: first, they needed to explain the mathematical and telescopic overturning of Aristotle's model. Second, they had to communicate—somehow—the mind-blowing immensity of what we now know lies beyond the Grecian's shattered spheres. Third, all of this had to be accessible to non-specialists, as well as to kids. In short, it was not an easy gig.

Wisely, they looked for a human face and a human story to carry the load—to bring these heavenly matters back down to earth, so to speak. Such a tactic, though, called for a careful choice of hero: a leader in logic; a master of mathematics; an expert in experiment; a scientific superstar. But whom, from all of history, should they pick?

Well, the good news is that someone else had already thought this difficult question through, and had come up with the perfect candidate:

> First of those who should be mentioned with reverence as beginning to develop again that current of Greek thought which the system drawn from our sacred books by the fathers and doctors of the Church had interrupted for more than a thousand years, was Giordano Bruno.[4]

The bad news is that this someone else was none other than Andrew Dickson White—the passage above is from his notoriously inaccurate *Warfare*.

And *Cosmos*—probably unknowingly—followed its advice.

## This Great Ecclesiastical Crime

In the Dark Ages, *Cosmos* tells us, space was pathetically parochial. The Earth was all that really mattered—"we were the center of a little universe," its script says. Then Bruno (1548–1600) came along. "There was only one man on the whole planet," trumpets Tyson, "who envisioned an infinitely grander cosmos." Yet "this was a time when there was no freedom of thought in Italy," we are warned.[5]

As the episode unfolds, we watch Bruno try to persuade his colleagues that the sun is just one of many stars, and that these other stars also have planets. We watch him exhorting his fellow intellectuals, lovingly and logically, to embrace a universe far greater than their small minds have so far imagined. We watch him be ignored, or laughed at, or attacked, or dismissed.

Unable to find acceptance anywhere across Europe, Bruno returns home—and is thrown into an Italian jail by the "thought police."[6] Then, in 1600, after nearly a decade of confinement, he is burned alive for his ideas. Tyson drives the key point home: "Why was the Church willing to go to such lengths to torment Bruno? What were they afraid of? If Bruno was right, then the sacred books and the authority of the Church would be open to question."[7]

With the lengthy animated tale now over, *Cosmos* switches to a computerized sweep across our solar system and into the universe beyond it. "Bruno had been right all along," Tyson confirms, with despair.[8]

A century and a half earlier, a certain John William Draper had been similarly outraged at Bruno's eventual fate:

> No accuser, no witness, no advocate is present, but the familiars of the Holy Office, clad in black, are stealthily moving about. The tormentors and the rack are in the vaults below. He is simply told that he has brought upon himself strong suspicions of heresy, since he has said that there are other worlds than ours . . . perhaps the day approaches when posterity will offer an expiation for this great ecclesiastical crime.[9]

Bruno, as portrayed in *Cosmos*, *Conflict*, and *Warfare*, is a conflict thesis martyr. His mortal sins were logic and science. His vision of millions of suns and millions of planets was unacceptable to the ever-dogmatic Catholics, who killed him for it. It was God versus science, plain and simple.

Except that it wasn't. Because they—Draper, White, and Tyson, that is—got the wrong guy.

## The Wrong Guy

The real Bruno, as it happens, was a wholehearted theist—he was deeply committed to the supernatural. In fact, one of his main arguments for a bigger universe was his belief in a bigger God. He wrote as much in his 1584 work *On the Infinite Universe and Worlds*:

So is the magnificence of God magnified, and the greatness of his kingdom made manifest: not merely glorified in one, but in innumerable suns; not in a single earth, one world, but in a thousand thousand, spoken in infinity.[10]

Indeed, it was not calculations based on observations that led Bruno to his apparently futuristic conclusion—it was his theology, philosophy, and mysticism. He believed that an infinite God was duty-bound to create an infinite universe—that He must do so by very definition. There was no divine choice in the matter; it was demanded by logic itself.

And yet, had Bruno done more homework, he might have known that earlier medieval thinkers had already analyzed this infinite-God-infinite-universe connection—and then rejected it. God, they had concluded, was entirely free to create whatever and however he wanted to. His universe, they said, might be infinite, and it might not.[11] Indeed, the only way to find out what he had actually done was to go out and make some observations—a notion which, as we shall see in Chapter 8, went on to give a significant boost to experimental science.

Bruno's most celebrated idea, then—the one that got him the *Cosmos* gig—was not so much ahead of his time, as it was behind it.

It gets worse for the Bruno-was-a-modern-scientist-stuck-in-the-wrong-age crowd: for the philosopher also believed that the Earth, stars, and planets were somehow alive. This is, incidentally, why he liked Nicolaus Copernicus's recent controversial ideas (published in 1543) about the Earth orbiting the sun—for a moving Earth fitted very nicely with his own conviction that it had some sort of life of its own.

Planets, he wrote, spun on their axes and orbited stars because they enjoyed the changes in temperature that came from doing so—they relished the variety of night and day, and of the seasons. Likewise, Bruno said, the sun and its inhabitants thrived on the Earth's supposedly radiated coolness: "In that hottest and light-giving of bodies [the Sun], there also live creatures which grow due to the refrigeration of the surrounding cold."[12]

None of these conclusions came, by the way, from looking through a telescope—for they did not exist in suitable form during Bruno's lifetime. He was therefore limited, as were his peers, to the twin powers of the naked eye and mathematics. But his math, when he tried using it, wasn't great; his published attempts often include basic geometrical errors, and are deeply flawed.[13]

Maurice A. Finocchiaro, a specialist on this period, comments on the "flimsiness of Bruno's scientific credentials" and explains that:

> His thinking contains a confusing mixture of metaphysical speculation, unorthodox theological criticism, and unconventional astronomy; an example is his animism, according to which natural bodies have souls, including the whole universe (whose soul is God) .... His manner of thinking is extremely obscure, exhibiting little reasoned argumentation.[14]

Bruno had strong words, however, for anyone who disagreed with his muddled thought: "Were we to ply such a man with further reasons and truths, it would be like ... washing the head of an ass: if you wash him a hundred times ... it is as if you'd never washed him at all."[15]

So Bruno, then, was not the man that the writers of *Cosmos* have depicted. A hot-headed and inventive mystical philosopher who was fascinated by Copernican astronomy? Yes. A coolly logical and brilliantly inductive mathematical scientist better suited to the labs of today? No.

But why was he killed?

## Thought Control on Astronomers

Bruno fell out with traditional doctrine early on in his career as a monk. He believed that Jesus was not God the Son, but a magician. The Trinity, he contended, was nonsense. The existence of the Church as an institution—the same Church he had once served—was, from his point of view, entirely unnecessary.

As he traveled, he continued to spread these ideas—and continued to insult anyone who questioned him along the way. In doing so, he built up quite the retinue of enemies. Finocchiaro says Bruno's writings "are full of satire, iconoclasm, blasphemy, profanity, sexual innuendo, and the like."[16] He was hardly trying to win friends and influence people; nor was he the puppy-dog-eyed victim presented in the *Cosmos* cartoon. And, in the end, the controversial combination of his personality and his philosophical theology led him to his death.

To be sure, none of Bruno's wild speculations or vicious attacks on his peers could ever merit the punishment he ended up receiving—it was both cruel in the extreme and wholly unjustifiable. And yet, the conflict thesis contention that he was murdered because of his scientific prescience is utterly false. While his ideas about infinite solar systems were indeed mentioned in the trial—which, contrary to Draper, had multiple witnesses present—it was not them which sent him to the stake.

We can know this last fact because many others before Bruno—including cardinals and other churchmen—had already suggested the possibilities of other worlds and of an infinite universe, and had faced no problems at all.[17] Historian Jole Shackleford, writing on Bruno, puts it simply: "The Catholic church did not impose thought control on astronomers."[18]

Interestingly, the *Cosmos* script actually seems to veer back toward the truth toward the end of its eleven-minute moral tale: "Bruno was no scientist," explains Tyson. "His vision of the cosmos was a lucky guess, because he had no evidence to support it."[19]

By then, though, it is really too late. The jibes have already gone in at the small-minded and illogical Church. The God-versus-science motif has already been emblazoned across the screen. Bruno has already been drawn soaring into space like a rational Christ, arms out wide in a faux-crucifixion scene. Bullying bishops have already spat, snarled, and shouted at our intellectual hero, like the villains in a Disney movie. One can't help but conclude that this is—like the exaggerated accounts of Hypatia in our last chapter—a piece of conflict thesis propaganda.

All of which prompts a question: aren't there better candidates out there for this role than Bruno? If one really wants to bash Christianity and big up science, why pick a heretical theologian—one who believed in an infinite God, taught about planets having souls, and messed up his math—as a role model?

Why, for instance, did the *Cosmos* scriptwriters not go for the far more obvious target of Galileo Galilei? Unlike Bruno, Galileo looked through a telescope, he conducted recognizably modern experiments, he performed multiple clever calculations, he came up with some physical laws—and, of course, he was placed on trial by the Church specifically for his astronomical writings. Wouldn't he have been a much better choice?

John Merriman, author of the definitive *A History of Modern Europe*, thinks so. When he was lecturing his history students at Yale on the dangers of "religious skepticism about any kind of rational belief," Merriman knew, straight away, which direction to point them in: "Look what happens to Galileo, who was lucky enough to have been burned at the stake by his friend the pope."[20]

But then that's not true, either.

## Illusions from the Devil

Here are the bare facts: Galileo Galilei (1564–1642), like Bruno, was a Copernican—he thought the sun was at the center of the solar system. When he began announcing it in public, he was warned off by Catholic naysayers, and was eventually banned in 1616—by the Inquisition, no less—from ever teaching the position again. In 1632, however, the talented astronomer-mathematician strayed once more; he wrote a scientific text advocating heliocentrism, and promptly ended up on trial for it. Threatened with imprisonment and torture if he refused to recant, Galileo stood his ground. He was found guilty of "vehement suspicion of heresy", was placed under house arrest, and his book was banned. He was never freed.

Let's be frank: it doesn't read well for the Church.

This story is, of course, in both Draper and White. Here's *Warfare*:

In vain did Galileo try to prove the existence of satellites by showing them to the doubters through his telescope: they either declared it impious to look, or, if they did look, denounced the satellites as illusions from the devil.[21]

And now *Conflict*:

On his knees, with his hand on the Bible, he was compelled to abjure and curse the doctrine of the movement of the earth. What a spectacle! This venerable man, the most illustrious of his age, forced by the threat of death to deny facts which his judges as well as himself knew to be true! He was then committed to prison, treated with remorseless severity during the remaining ten years of his life.[22]

It would appear, then, that Galileo would indeed have been a much better choice than Bruno for anyone wanting to push the conflict thesis. After all, religion has been caught, here, standing over science's body with dagger in hand.

But we have only heard half of the story.

## Scientific Folklore

Thomas Lessl is very interested in the art of storytelling; indeed, he is a professor of it. When it comes to the account of Galileo, he is a bit of an expert—he has carefully read through no fewer than forty popular retellings of it, and analyzed them to within a subclause of their lives. What he found is really rather fascinating—for Galileo and the Church, he says, has become a modern-day fairy tale.

In his paper "The Galileo Legend as Scientific Folklore," Lessl notes that important details are often skipped over lest they should spoil what can otherwise be a simple good-versus-evil storyline. Very rarely, for example, do books explain that Galileo was a devout Catholic. They don't mention that he was fully convinced of the deity of Jesus, and the resurrection, and the virgin birth, and that Scripture came "through the very mouth of the Holy Spirit."[23] They ignore that he argued forcefully for his faith on multiple occasions—often quoting Augustine, or Jerome, or Tertullian as allies. In fact, Galileo insisted, throughout his whole life, that: "The holy Bible can never speak untruth whenever its true meaning is understood."[24]

Other inconvenient truths are also jettisoned. Many cardinals and bishops, for instance, exalted the physicist, and backed his science publicly. On one occasion, when an angry sermon was preached against his heliocentrism, the offending minister was forced to apologize to Galileo, in print, by Catholic superiors.[25]

This pattern of censorship continues: the stubborn folk who refused to look through his telescope (White's "doubters") were actually secular philosophers committed to Aristotle, not members of the clergy. Indeed, at no point in the entire debacle were Galileo's scientific observations ever called into question by the Church—there was never any ecclesiastical doubt about his gathered data.

Two further comments are worth making (although many more could be added, including the awkward fact that Galileo had a Bruno-like tendency to mock other thinkers mercilessly, thus winning himself plenty of academic foes). The first is that Galileo did not have sufficient scientific evidence to verify his theory, for his results were incomplete and some of his deductions incorrect. As it happens, heliocentrism was not clearly supported by physical findings for another two centuries. The second is that Galileo was neither tortured nor jailed. Instead, he lived at a friend's house and then his own, with full permission to work on his physics. His game-changing treatise on motion, *Two New Sciences*, was written post-trial, and was published in 1638.

In papal Rome—the heartland of Catholicism—it sold out immediately.

By leaving these details out—a Christian Galileo with clerical allies and secular enemies; his failure to offer conclusive proof at trial when asked to; his continued freedom to publish and sell science books—the storytellers are misleading their audience. Allan Chapman, our doughty Oxford historian of science, puts it this way:

Let us remember that Galileo lived and died an obedient if disputatious son of the church, and that to see him as a martyr striving to make the world a cosy place for secularist ideologists to live in is pure mythology.[26]

Even though this has been pointed out again and again, Lessl doesn't think that the God-versus-science parable of Galileo will ever really go away—for there is too much hanging on it:

To expect the scientific culture to offer a more balanced view of these events might be reasonable, but it would also be unrealistic. Those who promulgate scientific folklore are not only naive historians but also partisan political actors . . . the values that are championed in the Galileo legend are shared with the broader culture of modernism . . . a belief in binary oppositions between reason and faith, knowledge and authority, and between Scripture and the light of nature.[27]

This is pessimistic, yes—but it may also be true. After all, if Yale's future leaders are being told by an eminent authority like Merriman that religion was up against reason in the early modern period, and that Galileo was killed by the pope, then the conflict thesis can only gain ground in the long run. Indeed, to help it do so, many in the Draper–White camp like to add a third super-famous name to those of Galileo and Bruno.

Who's that, then?

## The Myth Magnet

"I still can't see it," says Mike Brown, who is the incumbent Richard and Barbara Rosenberg Professor of Planetary Astronomy at the California Institute of Technology:

I go outside at night and watch the stars spinning around the north star . . . . But from where I stand and watch, it's all happening far over my head. I try to feel the earth spinning beneath my feet and to picture the globe rushing around the sun, but I just can't.[28]

This is some refreshing honesty. Brown, a professional astronomer who has known for all of his adult life that the Earth orbits the sun, admits he can neither "see" nor "feel" it happening. Imagining himself back in the sixteenth century, he says: "I like to think that I would have immediately recognized the simplicity and elegance of this idea and intuitively known it to be correct. But I suspect the opposite."[29]

Owen Gingerich is professor emeritus of astronomy and of the history of science at Harvard University as well as a senior astronomer at the Smithsonian Astrophysical Observatory—and he is in full agreement with Brown on how one would likely have responded to Copernicus's new sun-centered system: "You would no doubt have told him to get lost and to take all that nonsense with him . . . think how much harder it would have been to walk west than walk east! Totally ridiculous!"[30]

Nicolaus Copernicus (1473–1543), as it turns out, is quite the myth magnet. Brown and Gingerich have dealt with one example: that the Polish stargazer's decision to stick the sun in the middle was somehow obvious. Anyone who failed to accept his move at once, the assorted mythmakers suggest, must have been dumb, or stubborn, or dogmatic, or all three. Draper, for instance, says Copernicus "incontestably established the heliocentric theory."

In reality, though, his model was anything but obvious—it ran contrary to the physics of the day, and to human perception, and to common sense. In short, it seemed a bit daft. It wasn't even needed, either, since the age-old Aristotelian–Ptolemaic model had served astronomers, navigators, and farmers pretty well for two millennia. And, of course, heliocentrism also appeared to conflict with the simplest reading of Scripture—which leads us to a second Copernican myth.

Copernicus, so the story goes, was terrified of the Inquisition—after all, he was going against their Bible. To avoid likely torture at their hands, therefore, he only published when already on his deathbed. Here's *Warfare*, making the whole thing sound like quite the soap opera:

> To publish his thought as it had now developed was evidently dangerous . . . for more than thirty years it lay slumbering in the mind of Copernicus. . . . He dared not send it to Rome, for there were the rulers of the older Church ready to seize it; he dared not send it to Wittenberg, for there were the leaders of Protestantism no less hostile.[31]

Wrong, again.

In truth, the deeply Christian Copernicus was far more worried about being openly mocked by other natural philosophers than he was about being hurled into dungeons by inquisitors.[32] Why, for instance, didn't the atmosphere simply blow away as the Earth hurtled through space? Why didn't thrown objects get left behind as the planet moved on? At the time, his idea looked more than a little like madness—and Copernicus knew it.

Still, a few of his friends (including, incidentally, both a bishop and a cardinal), eventually managed to persuade him to publish, despite his fear of feeling foolish. One of the persuaders, the equally devout Andreas Osiander (1498–1552), added a placatory but not unreasonable preface claiming that the work was really only a mathematical one, and was best used for calculations only. (Bruno—ever the diplomat—later called Osiander an "ignorant and presumptuous jackass" for doing so.[33]) Sadly, Copernicus fell ill during the publication process, and only received the first edition as he died. And, just to be clear, he did not get sick and die on purpose.

Which brings us to myth number three: that the book was immediately banned. It wasn't. The initial response was mild interest from those in the know, a few raised eyebrows, and some understandable ridicule. Concerns were indeed raised by some in the Church, who felt uncomfortable with what seemed like a challenge to the Bible. Because of this, the work was "suspended until corrected." But this move came a full seventy years after publication, and it was no ban. Copernicus expert Robert S. Westman explains: "Catholics could buy and read the books, but they were to know that the doctrines contained therein were false, and they were instructed on which lines to expurgate."[34]

So then: heliocentrism was not obvious; Copernicus was not frightened of religious persecution; he did not deliberately wait until he was nearly dead to publish; his book was not banned. Is our magnetic Pole now myth-free? Well, no, not quite.

Andrew Dickson White isn't done yet.

## Three Myths, One Paragraph

Here, then, is *Warfare* again—and this time, displaying remarkable efficiency, White manages to squeeze three further questionable Copernicanisms into a single paragraph:

Calvin took the lead, in his *Commentary on Genesis*, by condemning all who asserted that the earth is not at the center of the universe. He clinched the matter by the usual reference to the first verse of the ninety-third Psalm, and asked, "Who will venture to place the authority of Copernicus above that of the Holy Spirit?"[35]

Let's take the Calvin myth first. A hugely influential theologian, John Calvin (1509–1564) was not afraid of a strong opinion or two—so *Warfare's* dramatic assertion seems feasible. But Calvin never actually wrote the words White ascribes to him—in fact, in all of his pages and pages of writings, he never mentions Copernicus at all.[36]

White's footnote for the imagined "quote" credits English cleric Frederic W. Farrar (1831–1903); but then the trail runs dry, for Farrar does not say where he got it from.[37] Once it turned up in *Warfare*, however, it was elevated to conflict thesis Scripture. The line has been used hundreds of times since, including by the eminent philosopher-mathematician Bertrand Russell (1872–1970)—bestselling author of *Why I Am Not a Christian*.[38]

Secondly, White's passage implies that Calvin the dogmatist thought science was at war with God himself. Yet here is Calvin, in one of his many books of sermons, on stargazing:

The study of astronomy not only gives pleasure but is also extremely useful. And no one can deny that it admirably reveals the wisdom of God. Therefore, clever men who expend their labor upon it are to be praised and those who have ability and leisure ought not to neglect work of that kind.[39]

These are hardly the words of a science-denier. So far, then, White's paragraph has erred badly on two counts. His third myth, however, is more subtle, and more deeply hidden—and it warrants some careful discussion.

Copernicus, White says, removed us from the "centre of the universe"—and, in doing so, it called dogmatic Christianity into question. How so? Well, because our eminent central position was supposed to be a sign of our special creation by God—and, in turn, that the Genesis account was true, and that mankind was the pinnacle of all of God's handiwork. And yet here was brave science, following the actual evidence—only to discover that we are not all that special, after all.

This Copernicus-demoted-humanity riff does the rounds in almost all modern retellings of the revolution, and is often accompanied by the and-in-doing-so-contradicted-Christianity bassline. Here, for example, is the famous New York University mathematician Morris Kline repeating it (alongside some other old favorites):

> A mathematician thinking only in terms of mathematics and unencum-bered by non-mathematical principles would not hesitate to accept at once the Copernican simplification . . . those who were guided chiefly or entirely by religious or metaphysical principles would not . . . a heliocentric theory that downgraded humanity's importance in the universe met severe con-demnation . . . John Calvin thundered "Who will venture to place the au-thority of Copernicus above that of the Holy Spirit?"[40]

Kline's full account of the events, as it happens, is one of the more balanced offerings available in the mainstream. The worst offenders—the most inac-curate, and the most partisan—often turn out to be school and university coursebooks, which is rather disturbing. American historian of science Michael Keas carried out a survey of a staggering 130 student guides, and found it in most of them: "Seventy-one percent of astronomy textbooks used in classrooms today perpetuate the Copernican demotion myth—sometimes laced with ex-plicit (though unwarranted) philosophical lessons bashing theism."[41]

The whole thing, though, is backward—for being at the center of the pre-Copernican universe had always been seen as a bad thing, not a good thing. Ever since the ancient Greeks—earlier, even—it had been the heavens that were considered to be perfect, and unspoiled, and aspirational. In other words, the further out one went from Earth, the more wonderful everything became. The center, by contrast, was where one found all the horrible stuff.

The fallen Earth of Christianity, at the hub of creation, was correspond-ingly considered to be corrupt and cursed. And, as if to emphasize the point, hell was even more central than that—for it was located in the middle of the Earth. When Galileo realized he could shift the Earth away from the center of the cosmos, he was thrilled—in fact, he saw it as a huge promotion for man-kind. He rejoiced, writing that we were no longer living on "the sump where the universe's filth and ephemera collect."[42] The heliocentric model, then, did not demote humanity—and it did not call into question how special we all are to God.

White, therefore, was mistaken on all three counts.

## An Antagonist Culture

Conflict thesis myths and heliocentrism, it seems, go hand-in-hand. The trio of Copernicus, Bruno, and Galileo frequently find themselves caught up in Draper and White–inspired fables, carrying out the bidding of whichever author feels they can make use of them. Lessl, our story analyzer, sees this as entirely understandable:

> Such narratives counterpoise the norms which scientific writers wish to attribute to their own culture against the attributes of an antagonist culture—which typically happens to be institutional religion. The presumed irrationality, credulity, and intellectual self-interest attributed to Galileo's opponents in the Church appear in these folk narratives as inversions of the rationalism, skepticism, and disinterestedness of science.[43]

In other words, the heliocentrism myths provide a simple "folk narrative"—one that helps to support a larger and grander idea: that science is rational, and religion is irrational.

Biologist Jerry Coyne, for instance, believes just that—in Chapter 1, we might remember, he questioned whether Christian physicist Tom McLeish should ever have been allowed to hold a Royal Society post.[44] Writing on the subject of the conflict thesis, he says:

> The toolkit of science, based on reason and empirical study, is reliable, while that of religion—including faith, dogma, and revelation—is unreliable . . . religion and science are engaged in a kind of war: a war of understanding, a war about whether we should have good reasons for what we accept as true.[45]

The name of his book gives his overall position: it is *Faith vs. Fact*.

But perhaps Coyne has missed something. Perhaps there is more to religion—and, for that matter, more to science—than he realizes.

## Fides Quaerens Intellectum

Is Christendom guilty of the charges that Coyne and Draper and White have thrown at it? Is it inherently irrational? Has it rejected reason? Has it looked down on logic? Has it eschewed experiment? How have the most influential thinkers in the history of the Church tended to approach the multi-faceted world of knowledge?

Well, in our previous chapter on the not-so-Dark Ages, we have already seen how the early Church embraced and encouraged ancient philosophy and literature, and how medieval monks gloried in math and science. To properly address the concerns raised by Coyne and others, though, this is not enough—for their accusations are aimed not only at Christian attitudes toward these other matters, but also toward the faith itself. So, what we really want to know is this: did these assorted thinkers switch their brains back off again when it came to the topic of God?

St. Anselm (c. 1033–1109) can help us toward an answer—his motto was *fides quaerens intellectum*, or "faith seeking understanding." Yes, Anselm says, Christianity is true—but, having assented to this, we should actively engage our minds to find out *why* it is true.

Dismissing any faith that "merely believes what it ought to believe" as "dead,"[46] he elaborates:

> There may be someone who, as a result of not hearing or of not believing, is ignorant of the one Nature [God], highest of all existing things. . . . And he may also be ignorant of the many other things which we [Christians] necessarily believe about God and His creatures. If so, then I think that in great part he can persuade himself of these matters merely by reason alone—if he is of even average intelligence.[47]

This is worth noting—Christianity can be shown to be true, Anselm believes, "by reason alone." He therefore set out, in his *Proslogion*, to offer rational arguments for Christianity—including his world-famous ontological argument, which remains the topic of hot debate among the best philosophers of today.[48] Logic and reason, Anselm taught, were to be both celebrated and used—after all, they were tools given to us by our God.

Anselm's approach, it is worth noting, was neither unique nor new. Cassiodorus (c. 485–585), a Christian scholar and educationalist of great repute, had championed active use of the intellect to meditate on Scripture five hundred years earlier—as historian Richard Raiswell explains:

> As Cassiodorus pointed out, the understanding of sacred literature is greatly enhanced if the reader approaches the page with a sound knowledge

of the rules of grammar, rhetoric, dialectic (i.e. formal logic), arithmetic, geometry, and music . . . [and so] they came to form the backbone of the medieval educational system.[49]

In fact, Christendom's highest-profile authorities, throughout its entire history, consistently promoted a fusion of faith with reason. Raiswell goes so far as to describe the mindset as "hyper-rational"—and, if anything, Christian theologians were far more obsessed with logic than modern scientists like Coyne are today.

Take *Sic et Non* (*Yes and No*) by Peter Abelard (1079–1142), for example, which laid out no fewer than 158 yes-or-no questions about Christianity. Budding clerics were expected to answer each one carefully by analyzing excerpts from the Church Fathers. Abelard had deliberately chosen ambiguous passages for them to use—for the whole idea was that students would have to demonstrate the clear use of propositional logic in support of their solutions.

*Sic et Non* had no mark scheme—only questions. Potential priests were judged on the quality of their reasoning, no matter which side of an issue they eventually landed on. Abelard emphasized this: "by doubting we come to examine, and by examining we reach the truth." His book went on to become a mainstay of medieval schools across Europe for centuries. Its queries included "Must human faith be completed by reason, or not?"; "Is God the cause and initiator of evil, or not?"; "Is God a single unitary being, or not?"— and even such mindbenders as "Does God's foreknowledge determine outcomes, or not?"[50]

From the early and then medieval Church, we can jump to the post-Enlightenment period—and, just as before, we find Christians using their brains to work through their faith.

Thomas Bayes (1701–1761), for instance, is one of the greatest logical analysts of all time—his work on probability now dominates modern math and physics. He was also a Presbyterian minister. So, did his logic and faith occupy different parts of his head, staying well apart from one another, as per Coyne's toolkits? Hardly. Instead, he made use of his discoveries to write a logic-driven paper entitled *Divine Benevolence: Or, an Attempt to Prove that the Principal End of the Divine Providence and Government Is the Happiness of His Creatures*.[51]

Few in the mathematical sciences have ever surpassed Bayes in terms of either originality and influence—but one man who perhaps has done is the extraordinary Austrian Kurt Gödel (1906–1978). His early work essentially tore the heart out of math, and ushered in a new era for the subject altogether—his paper on "undecidability" remains one of the most important ever written, and spawns new research even today.[52]

Gödel was a committed Christian and—like Cassiodorus, Anselm, Abelard, and Bayes—he applied his logic to his faith. At one point, he even wrote a formal symbolic proof for the existence of God. Its last line is: "Theorem 1. (A1), (A2), (A3) prove necessary actual existence of a godlike being, i.e. $\square$ ($\exists^{E}u$) G(u)."[53]

Huh.

Religious academics, then, do not switch off their minds when "doing God." They don't throw down one set of tools to pick up another. In fact, a Philpapers survey in 2009 found that over 70% of professional philosophers of religion—people who, by definition, use logic and reasoning to analyze beliefs—are themselves theists.[54] If Coyne was right that believing folk do not subject their faith to any kind of formal, rational study, then we would expect this figure to be rather lower than it is. In fact, we would expect it to be zero.

And, far from being a woolly crowd, these philosophers of religion can pack a propositional punch. Philosopher Quentin Smith—himself a naturalistic atheist who has worked extensively on physics—reckons that:

> If each naturalist [someone who rejects the supernatural] who does not specialize in the philosophy of religion (i.e., over ninety-nine percent of naturalists) were locked in a room with theists who do specialize in the philosophy of religion, and if the ensuing debates were refereed by a naturalist who had a specialization in the philosophy of religion . . . I expect the most probable outcome is that the naturalist, wanting to be a fair and objective referee, would have to conclude that the theists definitely had the upper hand in every single argument or debate.[55]

Huh.

Well then, there we have it: religion does do reason. To assert otherwise is to promulgate a myth. After all, the Bible itself says: "Get wisdom. Though it cost you all you have, get understanding" (Prov. 4:9).

Coyne, then, is wrong—or, at least, he is half wrong.

But is he also half right?

## Mythical Authority

So, if the original story was that religion is irrational and science is rational, we now know that we can throw out the first bit. Surely, though, the second bit holds by very definition, does it not?

Physics, chemistry, and biology—according to the Draper–White motif—follow strict protocols: observation, hypothesis, experiment, result. They allow no room for personal convictions, burnings in the bosom, unobserved assumptions, unchallengeable higher authorities, belief-driven biases, or hand-me-down creeds. Here, for example, is *Warfare*, praising the empiricist Roger Bacon (c. 1220–1294):

> It should be borne in mind that his *method* of investigation was even greater than its *results*. In an age when theological subtilizing was alone thought to give the title of scholar, he insisted on real reasoning and the aid of natural science by mathematics; in an age when experimenting was sure to cost a man his reputation, and was likely to cost him his life, he insisted on experimenting, and braved all its risks. Few greater men have lived.[56]

This is little short of an ode to the scientific method. Notice, in particular, that White contrasts "theological subtilizing" with "real reasoning"—just like Coyne.

Once again, however, *Warfare* has strayed into the mythological: White's claim that experimenting "was likely to cost him his life" is pure moonshine. One wouldn't have been able to move for empiricists (or, for that matter, alchemists) during Bacon's time. What's more, Ronald Numbers, writing on behalf of a host of the world's top historians of science, maintains that there is no record, anywhere or anywhen, of anyone of a Baconian persuasion being executed by any recognized church for either their thinking or practice: "No scientist, to our knowledge, ever lost his life because of his scientific views."[57]

The second myth in White's paragraph is harder to spot, but easy to feel: it is the notion that science is entirely dispassionate, and driven purely by fact and experiment; that it is unswayed, in any way, by human whimsy. In other words, he is helping to build the infamous myth of the scientific method.

Since Draper's and White's time, this myth has grown to be so ubiquitous and so inaccurate that many scientists have got thoroughly fed up with it. Here, for instance, is nuclear physicist Michael Brooks:

The brand identity of science is reinforced with adjectives such as logical, responsible, trustworthy, predictable, dependable, gentlemanly, straight, boring, unexciting, objective, rational. Not in thrall to passions or emotion. A safe pair of hands. In summary: inhuman. . . . We have been engaging with a caricature of science, not the real thing.[58]

Brooks is right. Any idea that the physical sciences are somehow able to operate outside of our humanity is highly misleading. Scientists, it should never be forgotten, are people too.

At times, then, some follow unjustified gut instincts; some hold unproven truths dogmatically; some worship their (academic) ancestors; some chase beauty over pragmatism; some fake results; some bend under financial or peer pressure; some misrepresent their work; some get things right by pure chance; some block the progress of others they envy or fear; and all of them experience pain, joy, frustration, and excitement—sometimes during the same experiment. Because, above all else, they are human.

And this humanity of science—despite its occasional shortcomings—is something to be celebrated, not denied. It connects science to other activities, to other struggles, to the arts and the humanities, to the broader community. Talking about science as it really is can make it attractive to people who might be left unmoved by the false notion of robotic lab workers and coolly detached theorists. Honesty is surely the way forward, is it not?

The aforementioned Tom McLeish, for example, is a genuine pioneer in the gloopy world of soft matter physics. His description of scientific discovery is rather different from White's:

A prevalent public view of the processes and methods of science sees cold logic, the application of a rulebook of experimental and theoretical practice and a disengaged, perhaps even dysfunctional, emotional approach by scientists to the subject . . . the real story is far more complex. . . . No scientific theory is born antelope fashion, fully formed in limb and energy, able to run for itself and keep out of harm's way. Our ideas emerge far more frequently as a marsupial birth—inadequate, vulnerable and powerless . . . they need to be loved into being.[59]

McLeish is not the only one trying to communicate to the non-scientist what science is really like from the inside. A plethora of works have arrived in recent years making the same point: that science—real science—is messy, and emotional, and much more like riding a rollercoaster than sitting on a train.[60] Yet still the fake painting-by-numbers view prevails—and many within science continue to push it. Why?

Historian of modern science Henry Cowles has written a whole book to answer this question. His answer is summed up in the following:

> There is no such thing as the scientific method, and there never was. And yet, "the scientific method" is alive and well. The idea of a set of steps that justifies science's authority has persisted in the face of constant denials of its existence. Why? Because "the scientific method" is a myth—and myths are powerful things. How we talk about science, how we account for its origins and argue for its results, instils mythical authority.[61]

Cowles, then, thinks a major motivation is the desire for "mythical authority."

Which is effectively another way of saying that science would like its own dogma.

## Flatly Asserted Dogma

Science utilizing or enforcing dogma? Is this really a thing?

It most certainly is.

Interestingly, Draper himself accidentally gives a brilliant example—for he preaches, with the full force of a fundamentalist, that "The day will never come when any one of the propositions of Euclid will be denied."[62]

One can almost picture him pronouncing this Ptolemaic psalm from a pulpit—with religious assent coming from the X Club choir. By "the propositions of Euclid," Draper meant the five ancient axioms of geometry, which had stood firm for 2000 years and which *Conflict* takes as obviously and scientifically true—they are articles of the rational faith, never to be doubted.

But, unbeknown to Draper, his contemporary Bernhard Riemann (1826–1866) did doubt them. A devout Christian and theologian, Riemann was prepared to stand against Draper's doctrine, and deny Euclid. When he did so, a whole new universe of math appeared in his equations, one that had been hidden from view for millennia. Just a few decades after Draper's death, Einstein would use Riemann's heretical work to formulate his world-changing theories of relativity.

Draper, the dogmatist, could never have dreamt of it.

Religious creeds also dominate modern quantum mechanics. For, while there is little argument about the underlying mathematics, its physical meaning has been hotly debated for the last century or so—and, as a result, the discipline has formed its own denominations.

The quantum-mechanical equivalent of Catholicism is the original Copenhagen interpretation, with Niels Bohr (1885–1962) as its St. Peter. The assorted Protestant groups include the many worlds interpretation, the pilot wave interpretation, and the transactional interpretation—all of which paint radically different pictures of reality.

So, how has rational science dealt with this in-house disagreement? Has it approached it with the cool, calm logic of scientific folklore?

Nope.

Here is theoretical physicist Sean Carroll, himself an evangelist of the many worlds church, bemoaning the behavior of the cardinals of Copenhagen:

> After a seminar in which another physicist explained Bohm's [pilot wave] ideas, Oppenheimer [Copenhagen] scoffed out loud, "If we cannot disprove Bohm, then we must agree to ignore him." John Bell [a waverer] . . . purposely hid his work from his colleagues . . . Hans Dieter Zeh [heterodox many worlds] . . . was warned by his mentor that working on this subject would destroy his academic career.[63]

Hugh Everett III, the first proposer of the many worlds doctrine, called the Copenhagen interpretation "flatly asserted dogma," and eventually left the discipline because of it.[64] Bell once asked in frustration about scientific censorship: "why is the pilot wave picture ignored in textbooks?"[65] The persecution got quite serious, as physicist and author Adam Becker points out: "When Zeh's students went looking for academic work, they were denied job after job, since they had not done 'real' physics. 'This', said Zeh, 'was something I will never be ready to forgive.' "[66]

Avi Loeb, the decorated Harvard astrophysicist, also describes his experience of falling foul of what he calls the 'scientific orthodoxy'. Having written a bestselling 2021 book that mooted the notion of alien visitation (in the form of a strange comet-like object called 'Oumuamua), he was largely written off by the astrophysical priesthood as a heretic. His response is telling:

> "I received numerous e-mails from astronomers, some tenured, who confessed that they agree with me but are afraid to speak out because of the potential repercussions to their careers."[67]

Becker, for his part, thinks that part of the problem here is actually a *lack* of logic among today's scientists: "Physicists are rarely trained in philosophy at all ... 'Philosophy is dead', declared [Stephen] Hawking ... according to Neil deGrasse Tyson, studying philosophy 'can really mess you up.' "[68]

"These are breathtakingly ignorant claims," Becker says.

And, even in the fast-paced, no-nonsense, engineering-on-the-edge world of Formula One motor-racing we find that Coyne's characterization is just too simplistic. Here is an anonymous aerodynamicist describing his day-to-day life on the job:

> You might believe there is no place in this world for dogma or blind faith. But you would be wrong ... some aspects of car design are taken as articles of faith, and the rest built upon them. ... We look to the demigods of aerodynamics for guidance, and uppermost in the current design pantheon is Adrian Newey ... the benefits of high rake are simply taken as gospel.[69]

So, then, there is indeed dogma and doctrine in science. And, perhaps more extraordinarily, there is even revelation, too.

The method for producing synthetic DNA, for instance, just popped into Kary Mullis's mind, fully formed, during an LSD trip. August Kekulé (1829–1896) literally dreamt up the structural rules of chemistry. Einstein got the idea for special relativity from an out-of-body transcendental vision.[70] That's three core principles in physics, chemistry, and biology, all coming from out of the blue—none of these folk were anywhere near a lab, and there was no hypothesis, experiment, or data in sight.

In the light of all this, it might be worth looking at Coyne's assertion once more: "The toolkit of science, based on reason and empirical study, is reliable, while that of religion—including faith, dogma, and revelation—is unreliable."[71]

Clearly, this statement is incomplete. For, as we have seen, science uses faith, dogma, and revelation. Religion uses reason and empirical study. Coyne needs to top up those toolkits.

## What Were They Up To?

We have covered a fair few conflict thesis myths in this chapter—each of which seemed convincing at first, and each of which fell apart on closer inspection. We have dropped in on Bruno, Galileo, Copernicus, Anselm,

Abelard, Bayes, Gödel, Bohr, Kekulé, and even a mysterious motor racing engineer. Story by story, Draper and White have been shown to be seriously wanting in their analysis. And yet still their narrative abounds.

And this, one must increasingly suspect, is because it is a convenient narrative for many. The likes of *Cosmos* and Coyne, and—from various preceding chapters—Dawkins, Freeman, Boorstin, Nixey, Sagan, Grayling, Tyndall, Huxley, Comte, and an assortment of modern textbooks are using mythology to either implicitly or explicitly call for science to somehow disarm, or dethrone, or dismiss, or displace, or defeat, or destroy religion.

The strange thing, though, is this: Draper and White themselves—the men who, for so many of the above, are spiritual leaders (even if they have never actually heard of them)—claimed to be faithful Christians, fighting on behalf of their faith. Both *Conflict* and *Warfare*, they said, were written to reconcile the hitherto estranged parties of science and religion—to save the marriage, so to speak.

Yet it would seem they have actually had precisely the opposite effect. Many of the commentators who insist on repeating Draper's and White's mistaken ideas are loudly advocating nothing short of a messy divorce.

So what on earth were these two men up to? And how did they get it so wrong?

# 7

# Bridges Badly Built

No Laufen Matter • Scarcely Recognized Even as a Class • Two Angry Men • A Friendship to Be Restored • The New Atlantis • A Thousand Threads • The Heterodox Side of Almost Every Question • Many Phases of Religious Belief • The Things That I Love to See • The Call of Your Innermost Nature • A Source of Strength to Both • Bridges Badly Built • Pure Materialism and Atheism • An Image Without a Soul

## No Laufen Matter

The engineers of Laufenburg, Switzerland, and Laufenburg, Germany, were feeling confident. They planned to connect their two towns by building a bridge over the river Rhein, and they were pretty sure that all would go well. After all, the span required was only around 100 yards, and the group had full access to an impressive range of twenty-first-century tools and techniques—so what could possibly cause them any bother?

Well, this: that the Earth is not a sphere.

Don't panic—we are not about to resurrect Lactantius and Cosmas. Instead, the point is that the Earth is not a *perfect* sphere. This makes deciding what is meant by "sea level" more complicated than it might at first seem; our planet's slight oblateness means the height of the sea actually varies around the globe.

There are, therefore, different sea level options to pick from—Germany uses that of the North Sea, while Switzerland has opted for the Mediterranean. The Swiss scientists, then, had a disagreement of a worrying 27 centimeters with their Teutonic teammates.

Oh dear.

Thankfully, though, we are talking about an expensive project and some very smart people—and so this discrepancy was spotted in advance. The architects took it on board and made the necessary adjustments: the troublesome figure was carefully added onto one side of the schematics.

Which, as it later turned out, was the wrong side.

The resulting structure—to the assorted embarrassment, exasperation, and expense of many—missed its mark by twice as much as it would have done without the change: a whopping 54 centimeters.

Oh dear, oh dear.

There is a lesson in all this for Messrs. Draper and White. For, as far as they were concerned, the territories of science and religion had a yawning gap between them—one that both thinkers desired to bridge. But how should they ensure, as all good project managers would, that the two sides would meet in the middle? Should religion be realigned? Should science be shifted? And in which direction should the chosen side move?

After much consideration and study, each man wrote up his bridging blueprint: *Conflict* was Draper's solution, and *Warfare* was White's. So, would these texts help others to cross the divide?

Or, like the Laufenburg laughing stock, would they make matters doubly worse?

## Scarcely Recognized Even as a Class

Given the significance of Draper and White to the emergence of the conflict thesis, it is hardly surprising that others have already wondered what made the two men tick. A handful of interesting theories have emerged as the front-runners, and it is worth having a quick look at them now. As we do so,

Draper and White and their world will hopefully become all the more real to us—for the pair of polemic penmen were, of course, *people*. So, let's put some flesh onto their nineteenth-century bones.

Perhaps the best known of all the conflict thesis origin stories comes from the late Yale intellectual historian Frank M. Turner. According to Turner, the still-fairly-new-on-the-scene professional scientific community was desperate to achieve genuine recognition as paid and specialized experts in their own right.

Up until this period—around 1850 or so—Turner explains that most scientists had been interested amateurs or hobbyists, and often members of the clergy. Now, however, science was morphing into a far more serious enterprise—and some of those more engaged in its practice quite fancied it as an honored form of employment in and of itself.

The continued presence of talented amateurs in the field, though, was an inconvenience for their cause. What made a full-time and professional "physicist" any more worth listening to (or, for that matter, worth paying actual money to) than a clever and committed scientifically minded vicar, for example—especially if the vicar was also a fellow of the Royal Society, had written plenty of scientific papers, and enjoyed the respect and friendship of other experimenters?

Turner put it thus:

> The major characteristics of British science were amateurism, aristocratic patronage, minuscule government support, limited employment opportunities, and peripheral inclusion within the clerically dominated universities. . . . The Royal Society was little more than a fashionable club as befitted a normally amateur occupation of gentlemen. In 1851 [computing pioneer] Charles Babbage complained, "Science in England is not a profession: its cultivators are scarcely recognized even as a class."[1]

Those men—and it was usually men—who longed to be admired and esteemed as "scientists," then, needed to bring about a change: that of professionalization. In short, they needed a clever and forceful narrative that would somehow push all the hobbyists out of their disciplines, and leave them as the new kings of the hill.

To some, at least, the answer was a rather obvious one: since most of the scientific amateurs were either employed by the church or otherwise part of the religious community, any decent argument that religion hampered science would do the job quite nicely. Turner again: "The professionalizers were not content merely to note or to ridicule the intellectual problems of the clerical scientist. In some cases they set out to prove that no clergyman could be a genuine man of science."[2]

One such example of this is Francis Galton's (1822–1911) statistical survey of "English Men of Science," undertaken in 1872. Galton, having shown that churchmen were poorly represented among the most elite groupings of scientists, called for a new and alternative "scientific priesthood" to be established across the kingdom—one which would take the lead in education and, eventually, in all of the relevant concerns.

Yet Galton, one of the founding fathers of statistics, had fiddled the figures. He cherry-picked values that suited his agenda and defined a true "man of science" in such a way that only those thinkers that fit his own mold could ever be counted as one. Turner says Galton's study

> Effectively excluded both amateur aristocratic practitioners of science and the more notable of the clerical scientists . . . no matter what the quality of the work . . . or the number of scientific honors and offices achieved, those people had almost no impact on Galton's data.[3]

Such professionalization was well underway when Draper and White were writing their manuscripts, so it is only natural that their works find themselves caught up in this same story. But were *Conflict* and *Warfare* really about a power-grab, and about proper recognition? Were they really manifestos for job creation? Were they really penned to prevent pesky preachers from practicing paleobiology without pay?

Or were things, perhaps, a little more personal than that?

## Two Angry Men

The strongest language in *Conflict* is reserved for the Roman Catholic Church. It had, Draper said:

> exercised an autocratic tyranny over the intellect of Europe for more than a thousand years. . . . Catholicism, as a system for promoting the well-being of man, had plainly failed . . . it had left the masses of men submitted to its

influences, both as regards physical well-being and intellectual culture, in a condition far lower than what it ought to have been.[4]

He was not nearly as harsh toward Protestantism; it was clearly Rome which had really riled Draper, and many viewed his work as a relentless attack on the older Church. What, then, could have made the chemist-turned-historian so very angry? Was his merely an academic frustration—or was there something else altogether going on under the surface?

Donald Fleming wondered the same. He spent forty years in Harvard's history department—but, a decade before he moved there, he wrote an influential biography of Draper. While interviewing some members of the family for his project, Fleming happened across an interesting little tidbit of information—one that might just explain why *Conflict* was so anti-Catholic.

Draper, it turns out, had an unmarried sister named Elizabeth—she is described as "the family rebel," and Fleming recounts a story about her that is really rather horrible. It seems that Draper had allowed Elizabeth to live with him, his wife, and his six children, but that this was not met with gratitude on her part. When William, Draper's 8-year-old son, became ill and was dying, the boy would repeatedly cry out for a Protestant devotional book—his favorite title—which alone could bring him comfort. Yet it was nowhere to be found. Then, when William finally passed away, it turned out that Elizabeth had hidden it:

> After his death [Elizabeth] laid it on Draper's breakfast plate. He met this cool challenge by ordering her out of the house. Though he never forgave her, she passed a happy, unrepentant life as a Catholic convert in Bridgeport, Connecticut. Perhaps her religious leanings lay at the bottom of the whole incident. If so, the experience may have helped sour Draper on the Catholic church.[5]

Maybe, then, unbridled fury is our surprisingly simple answer: maybe Draper wrote an angry book because he was an angry man. Could the same idea apply to White?

Well, yes, it possibly could.

As we might recall from Chapter 1, White's near-lifelong dream had been to found his own independent and non-sectarian place of learning. When he was finally in the position to do so—alongside Ezra Cornell—the project was almost scuppered by pious men decrying their proposition as irreligious, as White forcefully relates:

> Opposition began at once. In the State Legislature it confronted us at every turn, and it was soon in full blaze throughout the State—from the good Protestant bishop who proclaimed that all professors should be in holy orders . . . to the zealous priest who published a charge that Goldwin Smith—a profoundly Christian scholar—had come to Cornell in order to inculcate the "infidelity of the *Westminster Review*"; and from the eminent divine who went from city to city, denouncing the "atheistic and pantheistic tendencies" of the proposed education, to the perfervid minister who informed a denominational synod that Agassiz, the last great opponent of Darwin, and a devout theist, was "preaching Darwinism and atheism" in the new institution.[6]

White was incensed. Draper had fallen out with Rome; now White was mad at the whole of Christendom. Why couldn't these interfering and backward clergy, he fulminated, stop meddling with his well-meaning and forward-thinking aspirations?

Here, then, is a straightforward explanation for the genesis of both *Conflict* and *Warfare*: hell hath no fury like a nineteenth-century American intellectual scorned.

So is that it? Did Draper and White write their books to accelerate the professionalization of science and, simultaneously, to gain revenge on mean-spirited relatives, obstinate detractors, and Christianity in general?

No. They didn't.

### A Friendship to Be Restored

Turner is undoubtedly correct about professionalization as a general movement, and also about professionalizers such as Galton deliberately undermining the notion of a clergyman-scientist. Indeed, such considerations probably appealed to both Draper and White—but, in the United States, there was no established Church, and so the dynamics were not quite the same as in England. Draper, in fact, had found it incredibly easy to get both employment and respect in Stateside science. In short, it was not a big enough issue to make either man put pen to paper.

Neither was Draper's row with his sister. The family tale of poor William's book was passed along secondhand at best, and can rank only slightly higher than hearsay—even Fleming himself warns that "there seems to be no surviving evidence on this score."[7] Draper definitely despised Catholicism, as *Conflict* clearly demonstrates; but still, that was not his final motivation.

Neither was the nasty experience with Cornell for White—or, at least, it couldn't be unless the man could somehow travel backward through time. Those run-ins didn't occur until the 1860s, but White had already formulated most of his thesis long before that—as we shall soon see.

In the harsh light of day, all these well-rehearsed and easy-to-understand explanations fail. They fail because each ends up making the same key mistake—they ignore, almost entirely, what Draper and White themselves actually said they were doing.

Here is White, for instance, at the conclusion of *Warfare*:

In the light thus obtained [from "modern science"] the sacred text has been transformed: out of the old chaos has come order. . . . Of all the sacred writings of the world, it shows us our own [the Bible] as the most beautiful and the most precious; exhibiting to us the most complete religious development to which humanity has attained, and holding before us the loftiest ideals which our race has known.[8]

and now Draper, at the end of *Conflict*:

For Catholicism to reconcile itself to Science, there are formidable, perhaps insuperable obstacles in the way. For Protestantism to achieve that great result there are not. In the one case there is a bitter, a mortal animosity to be overcome; in the other, a friendship, that misunderstandings have alienated, to be restored.[9]

What is this? A "friendship to be restored"? The Bible, under the light of science, becoming both "precious" and "beautiful"? These are hardly the words of angry men bent on burning bridges between science and religion. Indeed, they are not—for both Draper and White were hoping to build one.

With the benefit of hindsight, of course, we know that they failed—that they actually managed to do the opposite. But how, and why? To find out, we will need to examine their motives in a little more detail—what was really going on in their heads?

## The New Atlantis

Draper, it turns out, is the trickier of the two to analyze. His refusal to use footnotes—not just in *Conflict*, but in general—means that finding out who and what most influenced him requires quite a lot of detective work. Fortunately, there are enough clues lying around for us to put a decent case together. The trail first gets warm way back in the 1600s—and with the mighty figure of Sir Francis Bacon.

Bacon (1561–1626) was a huge fan of the Reformation—as far as he was concerned, the momentous split between the evils of Catholicism and his own beloved Protestantism had brought about a wonderful new age of en-lightenment: "When it pleased God to call the Church of Rome to ac-count. . . at one and the same time it was ordained by the Divine Providence, that there should attend withal a renovation and new spring of all other knowledge."[10]

This "spring of all other knowledge," he believed, included a novel(ish) kind of science—one driven by primarily experiment and then analyzed, if it was possible, mathematically. Although others had toyed with this sort of science before him—the now-famous fourteenth-century "Merton calculators" from Oxford, for example, had imagined how it might have been done—Bacon was almost certainly unaware of it. In the 1530s, the fashions in education had swung suddenly to the classics, and more recent scientific work was almost entirely purged by the universities to make room for Greek and Roman literature—meaning Bacon and his peers may never have seen any of it.[11]

Back to Bacon's view: he thought God had provided humankind with rev-elation in two forms—the Bible, and nature. Each of these was to be carefully studied, with the outcome being—if all was done correctly—a single, unified, truth. The natural philosopher (by which we effectively mean "scientist"), therefore, was doing God's work every bit as much as the biblical scholar was. They were, in effect, a team.

Although many before him had intimated the same sorts of ideas—the idea of reading both Scripture and nature to know God went back at least as far as Augustine—Bacon pulled it all together into a coherent vision. For the first time, something recognizably like "modern" science was emerging as its own full-blown enterprise.

Bacon wrote down his philosophy in the form of a story, *The New Atlantis*. In his highly metaphorical text, a group of Catholic sailors are blown off course and arrive at an island unspoiled by contact with Rome. The islanders, who are unencumbered by ecclesiastical authority, are Christians nonetheless, since they found a Bible long ago in their past.

They also turn out to be scientist-scholars—they love nature, and constantly investigate it in order to better understand and worship God. Before long, the sailors are won over to this older, purer form of Christianity, and are then encouraged by the faithful philosophers to set sail once again and share its Gospel—an exciting fusion of Scripture, philosophy, and rational, experimental science.[12]

Many, especially in England, were smitten with Bacon's vision, which was soon taken up by the nascent Royal Society—or, to give it its full name, the Royal Society of London for Improving Natural Knowledge. Protestantism, reason, and natural philosophy (by which we effectively mean "science") were seen as bedfellows. This was rational religion—the best of both worlds—at its finest.

Soon, however, this happy equilibrium would be challenged: for rational religion was about to evolve.

## A Thousand Threads

For many of those on the Protestant side of the reformational divide there was a sense of a hard-won freedom—of having thrown off the shackles of a heavier and more controlling type of thought, and of the empowerment of individual believers to think about their Christianity for themselves. This, along with the heady successes of the likes of the Royal Society—a realized New Atlantis that could boast discoveries such as gravitational theory, mathematical motion, the calculus, gas laws, and more as its achievements—made rational religion a real favorite of the intelligentsia.

Such theology had been, at its inception, firmly committed to Scripture. The original idea (as championed by giants such as Newton, Boyle, and Hooke) was that Christianity as a worldview could be investigated and supported by means of experiment, analysis, and induction—that both its message and its God were true and real, and could be bulwarked by the study of creation.

As time went on, however, the momentum shift toward "science"—although the name itself did not yet exist—began to have a negative impact on the standing of tradition and Scripture. Some of Bacon's and Newton's successors felt less prepared than their forebears to hang their hats on the writings of the Church Fathers or the ancient creeds, or even on divine revelation itself. Science just felt more solid to them than parables did.

In questioning or even rejecting dogma, though, these thinkers did not see themselves as abandoning true Christianity—in fact, more often than not, they believed that they were purifying it, or enhancing it, or even restoring it, somehow, to its original form.

William Whiston (1667–1752), for example, sat in the same mathematical chair as Newton had at Cambridge, and went as far as arguing for what he considered to be a literal interpretation of Scripture. Yet Whiston (a theologian and historian to boot) maintained that early Christian thinkers had corrupted the faith—and, in 1712, he published a five-volume work called *Primitive Christianity Revived* that purported to show how. In it, he dismissed the time-honored and universal doctrine of the Trinity, found himself labeled as a heretic, and quit the Anglican Church—but still thought of himself as a faithful Christian. His story, far from being unusual, is actually typical of his age.

The Reformation, it became clear, had given rise to a rather unexpected side effect: by shifting spiritual responsibility more toward the individual and away from the Church, it had produced a generation much more ready to take a sort of pick-and-mix approach to theology, à la Whiston. In fact, by the early to middle 1700s, scores of supposedly Protestant scientist-philosopher-theologians were essentially creating "religions" entirely of their own. Rational theology had blurred, for some, into the free-for-all that is liberal theology—and now it was each man or woman for themselves.

A number of these thinkers, then, chose a theology that vastly favored natural philosophy over revelation—ending up with a stripped-down, maximum-science-minimum-Bible "faith" that was distinctly uneasy about anything overly supernatural. Such a move left them with a supreme deity of some hopelessly undefined sort: a hazy, nebulous oneness who was far off and unknowable, but who could still just about be worshipped—albeit in a vague sort of way—by studying, and enjoying, his-or-its creation.

Such was the "God" of the wildly popular English deism that emerged in the seventeenth and eighteenth centuries with the likes of Charles Blount (1654–1693), Matthew Tindal (1657–1733), John Toland (1670–1722), Anthony Collins (1676–1729), and Thomas Chubb (1679–1747) leading the way. This group of thinkers denied miracles, doubted the Bible, and did away with doctrine—for they saw them all as irrational. They adopted the term "natural religion," and argued that theirs was the true and pure faith of Christ. This deism, in reality, was the logical outcome of a post-Baconian theology that had become very dependent on the book of nature and very liberal with the book of Scripture. The famed historian Peter Gay put it thus: "Liberal Anglicanism and the dawning deist Enlightenment were connected by a thousand threads."[13]

And so, two centuries after Luther's great Scriptural stand against Catholicism, a whole host of progressives and liberals who claimed the German monk as their champion had conspired to throw aside pretty much everything that he had fervently believed. Bacon's crew of newly Protestant sailors had mutinied, and had headed off instead to their own island—a territory located in the Sea of Pseudo-Christianity.

And, if we really want to understand Draper, then one of these mutinous mariners is probably worth dropping in on: namely, Joseph Priestley (1733–1804)—the discoverer of oxygen.

## The Heterodox Side of Almost Every Question

A fellow of the Royal Society, Priestley was also a philosopher and theologian—making him exactly the type of character to continue what we might be tempted to call the liberal Protestant scientific revolution. Although he began his religious life in strict Calvinist orthodoxy, Priestley repeatedly changed his mind on matters of faith—each time moving himself further and further away from the traditional conventions, creeds, and catechisms.

"Christ was [only] a man like ourselves," he pronounced; to worship him as God would be "idolatrous." Priestley, the dissenter, was defiant: he said he "saw reason to embrace what is generally called the heterodox side of almost every question."[14]

It might seem (and it certainly would to Luther, or Boyle, or Hooke) that Priestley was not really a Christian at all—after all, the divinity of Jesus is perhaps the central belief of Christianity. But the chemist simply did not see things this way. Instead, he claimed to "defend Christianity, and to free it from those corruptions which prevent its reception with philosophical and thinking persons."[15]

Priestley believed, as strange as it maybe sounds, that he was running the same race as Bacon, and even as Luther. In fact, one phrase from his unorthodox *Catechism, for Children, and Young Persons* (1767) is extraordinarily reminiscent of Bacon's own writings:

> It pleased God to bring about a reformation, which is going on, and, we hope, will go on, till our religion be, in all respects, as pure, and as efficacious to promote real goodness of heart and life as it was at the first.[16]

Let's sum Priestley up, for it will be helpful: he was an eminent scientist; he loved theology; he was a minister in the Unitarian Church; he rejected many of Christianity's core teachings, calling them Catholic "corruptions"; he hated dogma; he saw himself as returning Christianity to its original form; he sought a religion that was rational; he viewed Luther as an inspirational figure; he believed he was continuing the Reformation.

Priestley died in 1804, but his thought lived on. We know this because his philosophy and his theology were repeated, approvingly, in an *Introductory Lecture on Oxygen Gas*, first published in 1848.

The author of this lecture was John William Draper.

## Many Phases of Religious Belief

It might be very easy to picture Draper as a nineteenth-century Richard Dawkins, or Jerry Coyne—but he was, in fact, no atheist. He believed, for instance, that science directed the "mind of a philosopher to a perception of the laws upon what it pleases God to govern the universe." Bacon, no doubt, would have approved of such sentiment.

Draper also praised Luther, saying his "grand idea which had hitherto silently lain at the bottom of the whole movement . . . the right of individual judgment" was a wonderful thing, and claiming that the Reformation had "introduced a better rule of life, and made a great advance toward intellectual liberty."[17]

In reality, however, Draper was far more of an heir of Priestley than he was of either Luther or Bacon. Here, for instance, are his views on Scripture:

It is to be regretted that the Christian Church has burdened itself with the defense of these books, and voluntarily made itself answerable for their manifest contradictions and errors.... Let it be remembered that the exposure of the true character of these books has been made, not by captious enemies, but by pious and learned churchmen, some of them of the highest dignity.[18]

So Draper sees the Bible as full of errors, but also praises Christian piety—just as Priestley would have done. The similarities continue: Draper identifies as Protestant, and detests Catholicism; he treasures science and reason as the best guides to truth; he believes in a "God" and considers himself a Christian; he rejects nearly all traditional doctrine; he considers miracles to be myths and Scripture to be suspect; he wants to restore Christianity, using cool logic only, to its original and uncorrupted form.

In that *Introductory Lecture on Oxygen Gas*, in fact, Draper openly lauds Priestley: "We must not impute it to mental weakness, but rather to a pursuit of truth, that in succession [Priestley] passed through many phases of religious belief."[19]

Draper, then, had taken the Reformation and run with it. He had ended up with a religion of his own—one that denied much of the Bible, and its revealed God, and its miracles—but that he thought of as Christianity nonetheless.

Well, OK—we now know Draper a little better. But we still haven't answered the actual question: why did he write *Conflict*?

Draper, as we have already suggested, hoped to bridge science and religion. Getting them to line up, though, would require careful Laufenberg-like alterations to one side—and Draper believed that it was religion that needed to change.

Dogma had to go—and so did most of the supernatural claims of Scripture, and a divine Jesus, and any assertion about God that came only from revelation. Provided religion was prepared to ditch all this, it would get along with science just fine, thank you very much. There would be no mismatch any more.

But Draper was not daft; he knew this argument alone was not enough. For *Conflict* to be successful, he had to show that its message was needed in the first place. He had to prove that religion, in its more orthodox guise, was well out of line with science.

And so *Conflict* was born: a list of tales about how Rome or popes or priests or doctrine or dogma or Scripture had dealt devastating damage to science and scientists and rationality and reason. A compendium of flat Earths, of medical misdemeanors, of the Dark Ages, of Hypatia, of Bruno, of

Copernicus, of Galileo, and all the rest. When his readers were faced with all this, he thought, they would have to concede the key point: dogmatic religion was no good for anyone—and especially not for science.

That, then, is the story of a very clever man who wrote a book all about reconciliation; and called it *Conflict*; and, in so doing, managed to make matters much, much worse than they ever had been before.

But what about White, and *Warfare*?

## The Things That I Love to See

Andrew Dickson White, fortunately for us, loved using footnotes. He was also a prolific letter writer and diarist, and these scribblings are now publicly available through the Division of Rare and Manuscript Collections at his beloved Cornell University. White, then, is much easier to analyze than Draper—provided, of course, one is prepared to read through his extensive back-catalogue. So, is his story the same as that of his infamous partner in historical crime?

Well, at first it seems like it might be—for White was every bit as much a son of the Reformation as Draper. In fact, his interest in it bordered on an obsession: by the late 1880s he had built up a staggeringly large personal collection of more than 30,000 books on the topic.[20] Later in life, he donated them all to the library at Cornell, along with thousands of other manuscripts which had tickled his various fancies.

A second and related love affair in White's life was with Germany—specifically its history and its educational atmosphere. In 1855, during several transformational years of study in Berlin, he wrote of a visit to Wartburg Castle—the place where Luther himself had been given protection by Prince Frederick III during the most dangerous times in his religious revolution. White's diary entry says he:

> lingered for a long time at [Luther's] window, looking at the wild scenery at which *he* once looked. I can see that a great change has come over me in the things that I love to see and to linger over. At my first sight, seeing it was all castles and abbeys, regardless of their tenants in great measure. Now I ask more—ask for places where something has been done for the race, for men as well as monks, and of these man-monks Luther is chief.[21]

This moment of introspection is especially helpful to our understanding of who White really was. First, he is clearly not the hard-nosed and combative anti-theist that many have imagined him to be. Second, he obviously admires Luther's Protestant remodeling of Christianity, seeing the "man-monk" as a hero. Third, he appears to be a man rather more connected to his own emotions than Draper. Indeed, this last observation, as we shall soon see, turns out to be a pretty significant one.

While in Berlin, White had studied the thought of many German philosophers and theologians, including the highly influential pair of Gotthold Ephraim Lessing (1729–1781) and Friedrich Schleiermacher (1768–1834). For the previous half-century or so, Germany's educational establishments had been making great moves in the academic world, and many young men at White's time saw them as the places to be—rich and avant-garde American students often enjoyed a stint there, reveling in the new and daring ideas that percolated in the "free" universities of Berlin and elsewhere. Such institutions were not as closely linked to the Church as their predecessors had been—and they were, accordingly, much more willing to flirt with religious controversy.

White adored everything about this brave new world—eventually, it led to him founding Cornell on the same sort of non-sectarian ideals. In the meantime, though, what matters most to us is that Germany had inspired White toward his very own big-picture understanding of the world—and of science and religion in particular.

## The Call of Your Innermost Nature

It was Lessing who had first got White thinking—as in *really* thinking—about the ideas that lay behind *Warfare*. In a series of books and plays, the German dramatist had argued that genuine religiosity involved an ever-evolving progression toward greater love and understanding. Each age could claim to have taken steps forward, as could each faith. The Old Testament, for example, was an "elementary primer"; the New Testament, itself just another placeholder, pointed toward an eventual "time of perfection." Jesus was "the first reliable, practical teacher"—but no more than that.[22]

Lessing maintained that all faiths would, one day, lead to the truth—but that none of them was there yet. No creed or dogma, therefore, should ever be considered complete or final; the idealistic "religion of Christ," which he believed was in keeping with "true religion," was not the same as that of the Bible, since the Bible was simply an ancient stepping stone to help earlier humankind along its way. White was fascinated by Lessing's arguments. And then, when he read Schleiermacher, he was entranced.

Formerly a Calvinist and a hospital chaplain, Schleiermacher the philosopher had turned his mind to almost every academic discipline at one point or another. In 1810, he helped found the University of Berlin that White studied at and, as such, became an object of the American's great curiosity. What White found in the German's writings appealed greatly to him. They offered the reader a new worldview altogether: one that the famed Reformed thinker Karl Barth described as a "theology of feeling."[23]

This theology was Schleiermacher's carefully devised tactic for dealing with the increasing amounts of religious skepticism that he saw around him, in Germany in particular. He considered himself to be a firm Protestant Christian—but understood that in the sense of Lessing's evolutionary faith, and not of Luther's far more orthodox stand on Scripture. It wasn't doctrine, or holy writings, or dogma, or revelation that mattered, Schleiermacher said, so don't worry about those things—instead, it was the inward witness of the heart that counts. Tackling the skeptics head-on, he challenged them to "become conscious of the call of your innermost nature and follow it."[24]

White was thrilled by both Schleiermacher's touchy-feely religion and Lessing's historical pilgrimage toward truth—so he combined the two to form his own take on life. The myths of the Bible, he decided, had indeed served a good purpose once upon a time—for they had helped simpler people to understand themselves, and to understand something of God, too.

Now, however, modernity was in need of a more enlightened analysis. Creeds, and doctrines, and literal approaches to Scripture were unhelpful in this new age—the real lessons to be learned from the Bible's fanciful stories were actually about personal morality, and love, and goodness.

God, White believed, had always gently and providentially revealed new ideas to humankind—to one era at a time. It was, therefore, our collective responsibility to grow as a species with each fresh revelation—holding on to the dogmas of a previous age was like crawling when we should be walking; it was crying for a mother's milk instead of eating solid food.

Here, then, is where Draper and White part ways—for, while Draper had advocated a return to a purer past to find an ancient truth, White was endorsing the exact opposite: he urged everyone onward into the future, in the hope of finding a newer and better one.

So where does science fit in to all this?

## A Source of Strength to Both

White was not himself a physicist, or a chemist, or a biologist—but that didn't mean, he thought, that his ideas were out of line with these disciplines. Indeed, *Warfare* described the new analysis of the Bible by the likes of Lessing and Schleiermacher as "science," and those carrying it out as "scientists"—as far as he was concerned, these thinkers were making discoveries every bit as important as those made in electromagnetism, or geology, or thermodynamics.

In fact, in a lengthy paragraph at the conclusion of *Warfare*, White seeks to draw his entire argument together. Look out, then, for the following themes: human progression as a lawlike principle; true religion really being about heartfelt goodness; traditional dogma as outdated and dangerous; the need for a new chapter for humankind; and, of course, for science as a guide:

> the Israelites, like other gifted peoples, rose gradually, through ghost worship, fetishism, and polytheism, to higher theological levels . . . their conceptions and statements regarding the God they worshipped became nobler and better . . . sciences are giving a new solution to those problems which dogmatic theology has so long labored in vain to solve . . . these sciences have established the fact that accounts formerly supposed to be special revelations to Jews and Christians are but repetitions of widespread legends . . . they have also begun to impress upon the intellect and conscience of the thinking world the fact that the religious and moral truths thus disengaged from the old masses of myth and legend are all the more venerable and authoritative, and that all individual or national life of any value must be vitalized by them.[25]

Here, then, is White's grand thesis: it is time for us all to grow up. God, via the sciences, has shown us a new set of truths, and we should respond accordingly. The older ways—the ways of doctrine, and Church Fathers, and a literal adherence to Scripture—should be left behind, for they are horribly unscientific, and belong to a previous and naïve age.

If we were only to listen to our hearts, White said, then we would appreciate the true meaning of the Bible, and grasp the true inner power of his vision. And, as an added bonus, science and religion would quite rightly be reconciled, once and for all:

> Thus, at last, out of the old conception of our Bible . . . has been gradually developed through the centuries, by the labors, sacrifices, and even the martyrdom of a long succession of men of God, the conception of it as a sacred literature—a growth only possible under that divine light which the various orbs of science have done so much to bring into the mind and heart and soul of man—a revelation, not of the Fall of Man, but of the Ascent of Man—an exposition, not of temporary dogmas and observances, but of the Eternal Law of Righteousness—the one upward path for individuals and for nations. No longer an oracle, good for the "lower orders" to accept, but to be quietly sneered at by "the enlightened"—no longer a fetich, whose defenders must be persecuters, or reconcilers, or "apologists"; but a most fruitful fact, *which religion and science may accept as a source of strength to both.*[26]

And so, with this aspirational image of the Bible drawing modern science and reimagined religion back together, *Warfare* draws to a close.

## Bridges Badly Built

Messrs. Draper and White, as we have discovered in this chapter, were intellectual engineers looking to build bridges. For too long, each man argued, religion and science had been violently estranged, and it was now time to reconnect them.

The problem, it was clear to both of them, was on the religious side of the divide. Draper saw the solution as obvious: religion, if it was to re-friend science, must become more detached and rational. White favored an entirely opposite fix: religion should become more personal and emotional. Where these opposite views shared a piece of common ground, however, was their stance on dogma—it needed to go.

And so they both wrote bestselling but error-packed books about how dogma had ruined scientific progress throughout Christendom—and, in so doing, they fooled the world.

That, then, is that: that is how *Conflict* and *Warfare*, works of intended reconciliation, came into being. And then, of course, it all went wrong—for, entirely against their authors' wishes, these two texts helped spark off a war that continues to this day. They are, to put it bluntly, badly built bridges. Yet neither of their authors was untrained, or untalented, or unintelligent—so how did they not see all this coming?

After all, plenty of their horrified readers did.

## Pure Materialism and Atheism

When Draper and then White's diatribes arrived on the scene, they provoked quite a reaction. Those who preferred their religion liberal—a loose, ill-defined, choose-the-bits-you-like approach to traditional ideas—were, broadly at least, fans.[27] One such reviewer at the *New York Evangelist* thought *Conflict* had succeeded in rationalizing and updating the faith, saying it was the "most powerful argument within my knowledge in defense of Protestantism and modern civilization."[28]

White received similar praise from similar outfits—but all this really tells us is that the two men had impressed those people who already agreed with their liberalism. Not everyone was so persuaded, however—and some of their detractors showed remarkable foresight in their criticism.

Essayist Orestes A. Brownson (1803–1876), a man who had done his fair share of exploring many different religious ideas before settling on Catholicism, thought that *Conflict*'s ancient-and-true-and-rational-and-natural faith was theological nonsense. Anyone denying revelation, and miracles, and doctrine, and a divine Christ, he opined, was hardly embracing a purer form of Christianity—instead, they were obliterating it: "[The] living and ever-present God, Creator, and upholder of the universe, finds no recognition in his physiological system . . . [it is] pure materialism and atheism."[29]

In other words, Brownson realized that Draper had made a very serious error of judgment. *Conflict* would not rescue Christianity from the skeptics, he warned—nor would it endow it with some sort of fresh intellectual rigor. The book's actual effect, he was convinced, would be to promote science as a viable alternative to religion. There would be no reconciliation; *Conflict* would set the two against one another.

And, before long, others would make the same points about *Warfare*.

## An Image Without a Soul

Andrew Dickson White received numerous letters from perceptive readers of his ideas, many of whom considered his reimagining of the Bible and spirituality to be hopeless and toothless, and no good for anyone. The physician John Shackelford, for instance, had some scathing words for him: "Christ's religion goes by the boards. A religion without the authority of God back of it is a religion without power."[30]

A family friend of White's, Mary Eaton, agreed with Shackelford. In an extraordinary and insightful series of letters, she spelled out to White exactly where *Warfare* had gone wrong—and also why it would almost certainly fail to achieve what he had hoped for it:

> If I am not mistaken, the object is to prove the Christian religion a "cunningly devised fable"; the Bible which Christians accept a tissue of falsehoods; its Divine author a myth; his son, that this same Bible pronounces "God manifest in the flesh," the greatest imposter the world has ever known; and then you inform us that it is true but in some high and mysterious sense![31]

White was trying to have his cake and eat it, Eaton says. He wanted to be seen as a pious Christian arguing for the truth of Christianity, yet also to deny any and all parts of that faith that he didn't personally like. This approach didn't wash well with Eaton. What he was offering would not draw people toward God, she wrote—if anything, it would push them away:

> What if you succeed in creating doubts in the minds of men, in taking from them all trust in the Revelation they have accepted as coming from God. What then? What will you give them in its stead? Your poor starving theories? . . . A religion evolved from human brains stripped of all that is Divine? An image without a soul?[32]

White wrote back, claiming that he was trying to "save the Bible"—to which Eaton amusingly replied "what do you suppose has saved it all these centuries without your helping hand?"[33] She went on: his enlightened Christianity, she said, was really no Christianity at all. White's ideas, Eaton predicted, were merely setting the stage for scientifically minded skeptics to jettison the divine once and for all:

You may not be an atheist in the sense that there is no God in the Universe, but when you declare that the Revelation he has made to man is fast crumbling away on account of what you call the "human theological foundations"—that Rock which he says the gates of hell shall not prevail against—I take the liberty by repeating: there is no God in your Bible.[34]

After two years of back-and-forth with Eaton, an exasperated White had had enough. He told the older woman that he would not be writing to her again. His parting shot was dismissive: "you know nothing whatever of the problem involved."[35]

But White was wrong. The likes of Eaton, Shackelford, and Brownson were onto something. For, like the project managers of Laufenburg, Draper and White had made careful adjustments to try and get the two sides of their bridge—science and religion—to meet up properly; and, just like those engineers, they had ended up making a horrible mess of things. They had wanted reconciliation—instead, they birthed the conflict thesis.

And yet, as impossible as it might seem, both *Conflict* and *Warfare* are plagued by an even greater irony than that. It turns out that when they went after Christian doctrine for being the ultimate enemy of science, they were engaging in friendly fire. For, in actual fact, no other body of thought has ever been of greater benefit to scientific thinking than the central tenets of traditional Christianity have—in the whole of human history.

As we are about to find out.

# 8

# Old Dogma, New Tricks

In the Beginning • The Deadly Foe of Scientific Inquiry • The Fire Ignited by a Faith Millennia Old • The Lord Our God, the Lord Is One • The God of Order and Not of Confusion • A Free and Unconditioned Creator • Principally in the Endeavor to Know Him • Fighting the Fall • Be in Some Part Repaired

## In the Beginning

"In the beginning," says the Bible, "God created the heavens and the earth."

The divine architect, we read, was rather pleased with his work: "God saw all that he had made, and it was very good." Later on, in the New Testament, we find that "in him all things were created: things in heaven and on earth, visible and invisible" and that "the universe was formed at God's command, so that what is seen was not made out of what was visible" (Gen 1; Col 1; Heb 11)

The Christian doctrine of creation is a richer one than many might have thought. It says that God built his cosmos entirely from scratch: *creatio ex nihilo*. It says he worked alone—"The Lord your God, the Lord is One"—there are no other gods. It says creation was both functional and delightful: "He has made everything beautiful in its time." It says the universe displays his power, his majesty, and his inventiveness: "The heavens declare the glory of God" (Deut 6; Eccles 3; Ps 19).

God may be one, but he is also three—and all the members of the godhead partook in the project. The wonderful culmination of their work—the masterpiece of the Father, Son, and Spirit—was humanity itself. "Let us make mankind in our image, in our likeness," said the Trinity, breathing life into Adam and Eve.

None of God's other creatures could claim the *imago dei*, or divine image—not even the vast host of angels who sang before his throne. Adam and Eve, thriving in the beauty of Eden, were magnificent. They were peerless on the Earth—except, of course, for those precious times when God himself walked alongside them in the garden.

"God blessed them," Genesis explains, "and said to them, 'Be fruitful.'" He imposed only a single rule: there was a tree that was out of bounds—"for when you eat from it you will certainly die" (Gen 1-2).

One rule, however, proved to be one too many. Humanity—proud, we-know-better humanity—rebelled against its perfect creator. This cataclysmic rejection of God resulted in a disharmony never before known to either heaven or Earth: "Cursed is the ground because of you," said God to his image bearers, "through painful toil you will eat food from it." Our ancestors and our world lay groaning under God's judgment—they, along with nature, were now subject to his curse (Gen 3).

The Fall of humanity had changed everything.

And then, through the darkness, shone the light of God's mercy: "I have loved you with an everlasting love," he declared to those who had joined the ever-growing rebellion. They could be won back, he said, and he himself would bring it about: "I will forgive their wickedness," God promised, "and will remember their sins no more" (Jer 31).

So the Trinity enacted a rescue plan: a plan that had always been in place, even before the foundation of the world. For God's intention in creation—from beyond the beginning—had been to demonstrate to his creatures the great depths of his "everlasting love." It had been for his children, having fallen into desperate trouble of their own making, to be forgiven, and saved, and restored. "They will be my people," he ordained, "and I will be their God" (Jer 32).

Yet their rebellion could not be left unpunished, for that would be an appalling cosmological injustice. God could not simply write off the Fall as if it had never happened. Here, then, was a cosmological tension: God is love; God is just.

The solution—the from-the-depths-of-eternity-planned solution—was divine self-sacrifice: God the Son would leave the glory of heaven, become human, and live out a life of suffering on the cursed Earth. He would walk, in the darkness and mess, among his broken people. Jesus, the anointed one, "made himself nothing, by taking the very nature of a servant, being made in human likeness" (Phil 2:7-9).

He would teach his fallen ones of the glorious image they bore—the *imago dei*—and of its almost unlimited potential. He would live the sinless life he had called them to live—even though he, in this incarnation, would be subject to all the same struggles and temptations as they were. He would love them, and love them perfectly.

With this first stage of his mission complete, the Son would then voluntarily take onto himself the guilt for humanity's Fall: "Greater love has no one than this," he told his disciples, than "to lay down one's life for one's friends" (John 15). He would be punished instead of them: he would die in their place. And, since his inherent worth was greater than that of all of creation, his sacrifice would be all-sufficient. The great price for the rebellion was paid.

Three days later, the Son was raised to a resurrection life by the Father, through the dynamic power of the Spirit—the Trinity was "making everything new." The wonderful invitation was issued worldwide: humankind could "be reconciled to God" (Rev 21; 2 Cor 5).

"A new and living way [has] opened for us," writes the jubilant author of the book of Hebrews, "let us draw near to God . . . having our hearts sprinkled to cleanse us from a guilty conscience and having our bodies washed with pure water." The death and resurrection of Jesus had fought back against the Fall, and there was now hope on the Earth again: "If anyone is in Christ, that person is a new creation. The old has gone, the new is here!" (Heb 10; 2 Cor 5).

The "new" came via the Holy Spirit, who would make his home in each individual believer: "He will guide you into all the truth," Jesus promised.

For God, as Spirit, is everywhere—he is *omnipresent*—so he could live in the hearts and minds of Christians, no matter where they were in the world. He would gently repair their broken consciences; he would tenderly restore them to their right minds. He would call them to live better lives (John 16).

This, then, is the grand story of Christianity. "All have sinned and fallen short of the glory of God," we are warned; yet God "demonstrates his own love for us in this: While we were still sinners, Christ died for us."

The headline is this: with God's help, the devastation of the Fall can be undone—one person, or behavior, or thought at a time. And, one day, God will take his welcomed-back creatures to a newly made and perfect heaven, for all eternity (Rom 3, 5).

"No longer" says the Apostle John about this idyllic future, "will there be any curse" (Rev 22).

## The Deadly Foe of Scientific Inquiry

Hang on a second—what is all this about prohibited trees, and divine curses, and rising bodies, and why, exactly, is it in a book about science? The ideas mentioned in the previous section seem about as far away from proper physics or correct chemistry as *the goop lab* was—if not further. After all, goop did, at the very least, manage to reference quantum mechanics, and frequency, and energy (albeit rather farcically). How, though, could the Holy Spirit ever help with hydrodynamics? Who needs to reflect on the resurrection before they can research resistivity? Isn't religious dogma—to put it simply—just about the polar opposite of real science?

Well, that was certainly the Draper and White line, as we know. *Conflict* declared emphatically that: "From the time of Newton to our own time, the divergence of science from the dogmas of the Church has continually increased."[1]

*Warfare* wholeheartedly agreed: "from the supremacy accorded to theology, we find resulting that tendency to dogmatism which has shown itself in all ages the deadly foe ... of scientific inquiry."[2]

Their claim that dogma and science are mortal enemies has become so commonplace within modern culture that it now seems rather obvious—a truism, almost. The Nobel Prize–winning chemist Harry Kroto, a fellow of the Royal Society, went all in on it when explaining why biologist Michael Reiss—a practicing Christian—was not fit to work for the organization:

Unfortunately Reiss, who is, apparently, a very nice guy, was in the wrong job. He, together with all religious people—whether they like it or not, whether they accept it or not—fall at the first hurdle of the main requirement for honest scientific discussion because they accept unfounded dogma as having fundamental significance.[3]

Kroto is not alone. Nowadays, we can easily find his Draper and White–inspired stance repeated in popular non-fiction books, TV documentaries, assorted school and university course texts, magazines, bestselling novels, YouTube videos, and every corner of the internet. There is one place, however, that we will never come across it: the academic literature of the history of science.

And that's because it is wrong.

## The Fire Ignited by a Faith Millennia Old

Friedrich Nietzsche (1844–1900) was, perhaps, the arch-atheist—no intellectual has ever attacked God with as much insight and self-consistency as the German philosopher did in the latter half of the nineteenth century. Interestingly, however, he did not take the same line as Kroto and his cronies. Instead, he carefully pointed out that even the very concept of "honest scientific discussion" actually owes its existence to Christianity:

> Strictly speaking, there is no such thing as science "without any presuppositions" . . . a philosophy, a "faith," must always be there first of all, so that science can acquire from it a direction, a meaning, a limit, a method, a right to exist . . . we men of knowledge of today, we godless men and anti-metaphysicians, we, too, still derive our flame from the fire ignited by a faith millennia old, the Christian faith.[4]

If Nietzsche is right, then this is a crushing blow to Draper, and White, and Kroto, and the rest of the gang. But is he? Was Christianity really the crucible in which modern science first began to glow? If so, how? How does the Bible provide science with its "presuppositions"? How does its account of creation, fall, loss, and redemption give science "direction," or "meaning," or "method"? How does Scripture grant science a "right to exist"?

Maybe it just doesn't—maybe Nietzsche is mistaken. The thing is, though, that he is backed up by a whole body of recent research in the field. Peter Harrison, for example, is one of the most respected historians of science and religion working today—and, in one passage about the formation of Kroto's Royal Society back in the mid-1600s, he writes:

> Whereas we often tend to think of religious influence manifesting itself unhelpfully in the content of scientific ideas, far more important is the manner in which religion lent social legitimacy to scientific activities and institutions, provided motivations for key individuals in those institutions and, not least, informed their goals and methods.[5]

What is actually going on here? If Christian doctrine inspired scientists—if it gave them their philosophical underpinnings, suggested their methods to them, legitimized their work, and drove them onward toward their findings—then how, exactly, did it do it?

## The Lord Our God, the Lord Is One

We shall begin with a simple and central claim of Christianity: that there is only one God.[6] This core belief—monotheism—may appear to have nothing to do with science, but it does. In fact, it might just be a vital ingredient of it. How so?

Well, a monotheistic worldview—in the vast majority of studied cases—has led its holders to expect both consistency and regularity in the natural world. This, in turn, has led to them going and looking for them—and, ultimately, to the discovery of laws in nature. But what, precisely, is the underlying cause of this strong monotheism–natural law link?

Here is one theory: societies or individuals who picture the world as governed by many gods can tend to think of their environment as the unpredictable product of supernatural chaos. Such a world is not open to rigorous and methodical investigation by us humans—for it is, instead, entirely subject to the whims and fancies of a host of inconsistent, inscrutable, or even capricious spirits. The true reasons behind a physical event, then, are unclear—and matters could easily play out very differently the next time. Science, in such a scenario, is a non-starter.

By contrast, monotheism gave rise to one of the cornerstones of modern science: the belief that an experiment conducted on a Monday will give results that are still usable on a Tuesday, since the same God is in total charge of his universe on both days. Natural philosophy, under such assumptions, becomes a viable activity—and one from which we might expect some meaningful findings.

Yet isn't there a hole in the logic here? Why should one supreme God mess around with nature any less than a panoply of lesser deities would? Couldn't a single omnipotent being cause just as much pandemonium—indeed, more—than a collection of lower gods might?

The answer, of course, is yes—but this is not the type of God that Christian monotheism has typically championed. Instead, the God of the Bible is one of order and of rationality; a ruler who breaks his imposed laws only in the most exceptional of circumstances. Historian of science Mark Worthing puts it this way:

> One might think that monotheism, especially those forms with strong affirmations of God's sovereignty, would have been little different to polytheism . . . didn't monotheism simply roll up the functions of the various competing deities into one package? But this is not what happened . . . in most cases belief in one God—especially an all-powerful, all-knowing sovereign God, led people to look increasingly to natural causes. An all-powerful, sovereign God, while certainly capable of intervention—indeed of miracle—would create and govern the world in such a way as to ordinarily follow regular and therefore comprehensible patterns.[7]

In practice, then, what a culture believes is going on "up there" can have a very real effect on what it believes is going on "down here." Christian monotheism, thanks to its accompanying notion of a sensible and consistent God, proved itself to be a good foundation for science—it "led people to look increasingly to natural causes." Here, then, is a case of a seemingly unscientific statement of faith actually being rather helpful to the rational and scientific thinker.

And it is not the only one.

## The God of Order and Not of Confusion

Christians, throughout history, have asserted that God made *everything*, and that he made it all out of *nothing*.[8] From this one point of dogma flow many secondary conclusions: that God was not compromised by having to use pre-existent material; that creation is God's own entirely bespoke design; that everything around us is the product of a perfect and rational mind; that the world—in its original form, at least—was carefully and meaningfully constructed and is, as God himself put it, *good*.

All of these considerations turn out to have positive implications for science—indeed, they have provided it with a warm and cozy environment in which to thrive. For, if nature is God's handiwork, and if God is reasonable, then his universe should be reasonable, too. And reasonable universes, of course, are both easier and more enjoyable to study.

What's more, the Bible says that God has made us humans in his "likeness"—that our minds are in some way similar to his. This is, scientifically speaking, also good news—for if our minds can be in step with the architect of the universe, it follows that we might be able to comprehend his blueprints. Science, then, is worth a shot: we, as God's special creatures, have at least a chance of comprehending our surroundings.

It was this line of thinking, for instance, that inspired Johannes Kepler toward his famous mathematical analysis of planetary orbits. Writing in 1604, he explained his underlying philosophy:

> For the theatre of the world is so ordered that there exist in it suitable signs by which human minds, likenesses of God, are not only invited to study the divine works, from which they may evaluate the Founder's goodness, but are also assisted in inquiring more deeply.[9]

Kepler attributed his astronomical successes to the doctrine of creation—he could only make his discoveries, he said, because God had made the world "ordered" and "suitable" for study, and had then given him a God-like mind which was, by its very nature, "invited" and "assisted" in exploring the cosmic structure. Newton was one of many others to agree with Kepler—he also felt his feats were wholly dependent on the character and conduct of his creator:

> The world, which to the naked eye exhibits the greatest variety of objects, appears very simple in its internall constitution when surveyed by a philosophic understanding. . . . It is the perfection of God's works that they are all done with the greatest simplicity. He is the God of order and not of confusion.[10]

The same can be said for the likes of Boyle, Hooke, Descartes, and nearly all of the main players in the so-called scientific revolution of the seventeenth century. These pioneering natural philosophers believed their created minds could uncover the laws of created nature precisely because God had designed the former to be discerning and the latter to be discernible.

Robert Boyle, for example, was perhaps the first recognizably "modern" chemist—but he argued that the reason he could predict particle motion was because God had set everything up for him to do so:

> It is intelligible to me that God should . . . impress determinate motions upon the parts of matter, and that . . . he should by his ordinary and general concourse maintain those powers which he gave the parts of matter to transmit their motion thus and thus to one another.[11]

Note that Boyle's God does not control matter arbitrarily, but in such a way that it behaves consistently—and also that Boyle believes this is "intelligible" to him. In such statements of faith are the seeds of modern science.

This doctrinal logic did not die out with the Kepler–Boyle–Newton generations. Two hundred years later, Michael Faraday and James Clerk Maxwell—two of the most important thinkers of all time, and the main developers of electromagnetic theory—were drawing on the same theological ideas as their predecessors. Both Faraday and Maxwell felt they were reading words that had been "written by the finger of God," explains MIT physicist Ian Hutchison. It would be nonsense, he says, to think that their religion somehow got in their way: "On the contrary, their spiritual beliefs were essential parts of the strength of character and the view of nature that empowered them to make their transformative contributions to science."[12]

Indeed, the self-same phenomenon is alive and well in the twenty-first century—for many of the very best practitioners today say that their science is both undergirded and driven by their religious beliefs. Oxford's John Lennox insists "it is the existence of a Creator that gives to science its fundamental intellectual justification."[13] Acclaimed engineer and computer scientist Rosalind Picard agrees: "the existence of simple orderly mechanisms are not only consistent with God's nature, they are a reflection of it."[14] Eminent quantum physicist John Polkinghorne writes that because "we are creatures made in the divine image, then it is entirely understandable that there is an order in the universe that is deeply accessible to our minds."[15]

Theologian and literary historian Rebecca McLaughlin has considered all this in the round. She concludes that the doctrine of creation is effectively "the first hypothesis of modern scientists," and spells it out as follows: "If a rational God made the universe and endowed humans with an intelligence that echoed his own, perhaps his image-bearing creatures would be able to discern his laws."[16]

This long-established and well-attested relationship, however, seems to be entirely lost on both Draper and White. Somehow, *Conflict* manages to get the whole church-dogma-led-to-natural-law situation both upside-down and inside-out: "The Christian was convinced of incessant providential interventions; he believed that there was no such thing as law in the government of the world."[17]

*Warfare*'s account is less blunt, and dangerously subtle. A skilled diplomat, White used sleight of hand to transform Newton—a deeply religious thinker who had both espoused and relied upon traditional doctrine—into some sort of anti-dogmatic hero: "[Through] Newton, had come a vast new conception, destined to be fatal to the old theory of creation, for he had shown throughout the universe, in place of almighty caprice, all-pervading law."[18]

White, in order to serve his greater purpose of ridding religion of dogma, invented a battle between "all-pervading law" and the "old theory" of Scripture—when, in reality, one was derived from the other. He made them enemies when they were friends. Rather depressingly, his rhetorical trick proved to be highly effective. Nowadays, the supposed enmity between scientific law and Christian doctrine is taken by many as a given. As a piece of persuasive writing, *Warfare* has been extraordinarily successful. As an accurate record of the historical interplay between science and religion, however, it is sorely lacking.

So, in our continued effort to try and retrospectively right some of White's wrongs, we shall plough on—for the Church's teaching on creation had yet one more pro-science string to its bow, and one that brought about another intellectual leap forward. The key idea was this: things didn't have to be this way.

## A Free and Unconditioned Creator

Something rather strange happened in Paris back in AD 1277. Indeed, a whole series of odd events took place, many of which appear to play straight into the hands of overly enthusiastic conflict thesis advocates such as Gibbon, or Sagan, or Tyson. These incidents are worth more than a moment of our time, for they will lead us to some surprising conclusions—including the seemingly paradoxical discovery that Church dogma can, on occasion, be remarkably *un*dogmatic.

In thirteenth-century Europe, much of the academy was dominated by Aristotle. Recent reacquaintance with his 1500-year-old work in the West had resulted in the voracious study of it, and the famed theologian Thomas Aquinas was one of many thinkers flying the philosophical flag for his ancient counterpart. Indeed, the Grecian was fast becoming an unlikely darling of the Church.

It helped that Aristotle's position could, if viewed from a suitable angle, appear monotheistic, and that much of his thought dovetailed quite neatly with that of Christianity—or, at least, it would if interpreted in the right way. Aquinas and others adored Aristotle's logical and holistic worldview, and longed to harness it for Christian purposes. Reason was the thinking Christian's champion, the Aristotelians argued—reason would lead us to truth.

Not everyone was convinced, however, and underneath this apparent academic bliss there was the distinct rumble of division. For, as much as Aquinas and the gang had sought to blend Aristotle with Scripture and tradition, some of it really didn't fit. The Aristotelian cosmos, for example, was eternal in both directions, whereas the Bible described a clear beginning and a clear end. Other issues also rankled, such as the nature of the soul; suffice to say, there was still quite a bit to be washed and ironed out.

In France, though, the clergy were not all that keen to do Aquinas's laundry for him—instead, they voted to throw his garments aside and buy some nice new outfits to take their place. Edward Grant, a veritable giant in the academic world of the history of science, takes up the story:

> The Church tried initially to ban the natural philosophical works of Aristotle (in 1210). When that proved unsuccessful, there was an effort (in 1231) to delete the offensive parts of Aristotle's philosophy by censorship, but this was never carried out. Finally, in 1277, the bishop of Paris issued a condemnation of 219 articles. . . . Many, if not most, of the condemned articles were drawn from Aristotle's natural philosophy. Some twenty-seven articles condemned different versions of the eternity of the world.[19]

"Aha!" cries Jerry Coyne; "I told you so!" yells Catherine Nixey. For here, as plain as the nose on A. C. Grayling's face, we have dogmatic theology squeezing the life out of rational science—and destroying classical thought as it does so.

But not so fast—for that isn't how matters played out.

Perhaps we should let Grant continue:

The most significant aspect of the Condemnation of 1277 was not the condemnation of Aristotle's offensive ideas about Christian doctrine, but rather the condemnation of his ideas that seemed to limit God's absolute power to do whatever he pleased short of a logical contradiction. . . . They firmly believed that for God Aristotle's natural impossibilities were all "supernaturally possible."[20]

Well, this is interesting. What Grant is saying is that it was Aristotle who was overly dogmatic, and that the Church was being far more open-minded. Aristotle had declared some occurrences to just be downright impossible— even if they were allowable by logic. The churchmen of Paris didn't like this; they felt that all possibilities should be kept on the table. So, in a counter-strike, they banned those statements that banned those statements (read this twice). Here are some examples of what they got rid of:

34. That the first cause [God] could not make several worlds.

48. That God cannot be the cause of a new act, nor can he produce something anew.

49. That God could not move the heavens [or Earth] with a rectilinear motion; and the reason is that a vacuum would remain.

141. That God cannot make [a property] exist without a subject nor make several dimensions exist simultaneously.[21]

Under Aristotle, it was impossible for there to be more than one world-or-universe; or for any particle, force, or field to come into existence; or for the Earth to move through space; or for there to be multiple dimensions overlapping one another. Even if God willed any of these things, he said, they would still never happen. The Church—in Paris, at least—disagreed: God could do whatever he wanted to, they maintained, whatever Aristotle might think.

Grant explains the hugely beneficial long-term effect of this debate:

After 1277, [Christian philosophers] not only chose to imagine that all of Aristotle's natural impossibilities were possible, as well as others that he had never considered, but they assumed, hypothetically, that God had actually performed them . . . they entered the realm of "let's pretend" and began to discuss topics that were literally out of this world, which stirred their imaginations in remarkable ways. The major consequence of all this was that hypothetical and counterfactual discussions became a vital part of medieval natural philosophy. The medieval imagination was free to "probe and poke around". . . . Their achievements were destined to influence some of the major figures of the Scientific Revolution.[22]

Once again, then, we find that the truth is the precise opposite of the *Conflict-Warfare* fiction: Christian dogma, rather than putting the blinkers on thinkers, opened up their minds to new possibilities. It was, therefore, a significant source of genuine and playful freethought about nature.

Even now, the idea that God is wholly unconstrained plays its part in the philosophy of science. Donald Mackay (1922–1987), a decorated physicist and neuroscientist, said this profoundly freeing doctrine still benefits those working in laboratories today:

> The God of the Bible is no mere craftsman. . . . He is a free and unconditioned *Creator*: the giver of being to a world of His own devising whose nature could neither be defined nor fully deduced by reference to any first principles. . . . Biblical theism, by denying that we can lay down in advance what the world ought to be like, offers positive encouragement to the experimental approach to nature that we now take for granted as "scientific."[23]

The belief in a "free and unconditioned Creator" meant that scientists had to expand their mental horizons, and then check their ideas by experiment to find out what God had actually done. This aspect of the doctrine of creation, therefore, helped to initiate and then support the much-lauded scientific method—the same one which is now taught, as a matter of routine, to children worldwide. Yet how often, we might ask, are they ever taught about its philosophical and theological origin?

One suspects—thanks in no small part to Messrs. Draper and White—that the answer is "rarely."

## Principally in the Endeavor to Know Him

One of those who most firmly asserted that God could create entirely as he pleased was the mighty philosopher and mathematician, René Descartes (1596–1650). The Frenchman's belief in God's intellectual freedom led him, in turn, to his own: Descartes was happy to challenge previous thinkers such as Aristotle, and even more so to reject them. As a result, he came up with several brand new ideas of his own—including his famous proof of his own existence: "I think, therefore I am."[24]

What is of most interest to us right now, however, is the motivation behind his work—for *why* did Descartes bother to think in the first place? Well, as a devout Catholic, he gives his reason plainly:

> I think that all those to whom God has given the use of this reason have an obligation to employ it principally in the endeavour to know him and to know themselves. That is the task with which I began my studies; and I can say that I would not have been able to discover the foundations of physics if I had not looked for them along that road.[25]

In other words, Descartes wanted to study the world so that he could have a better relationship with God. This is in keeping with the idea that God had given humanity two books by which to know him: the book of Scripture, and the book of nature. Such a view was espoused by Bacon, Newton, Boyle, Hooke, and, if we are being honest, pretty much every natural philosopher of the period—as we saw in our previous chapter. The fundamental idea can be traced back at least as far as the New Testament, in which St. Paul wrote to the Romans: "Since the creation of the world God's invisible qualities—his eternal power and divine nature—have been clearly seen, being understood from what has been made" (Rom. 1:20).

Margaret Osler (1942–2010) built a career analyzing the thinking, motivations, and intentions of the big hitters of the scientific revolution—perhaps no one knows what made them tick better than she did. Her conclusions are striking:

> The Bible was a powerful source of teleological thinking. According to the Old Testament tradition, God created the world *ex nihilo* by his power and will . . . his wisdom and power are evident in the creation. A biblical verse frequently quoted by early modern natural philosophers proclaims, "The heavens declare the glory of God; and the firmament sheweth his handywork."[26]

Why did these "early modern natural philosophers" study the world? Because they wanted to know the God who had made it—and, in turn, they wanted to make him known to others, so that they might be blessed by the knowledge themselves. Robert Boyle, again, is a good example: in his will, he left money to fund a series of scientific lectures with the specific purpose of "proveing the Christian religion" to those who were currently unconvinced. These lectures, incidentally, still run today, and are given by some of the biggest names in the business.[27]

So far, then, dogma seems to be far more of a blessing to science than either Draper or White would have us believe. But what of the more heavily theological concepts of the Fall, or of original sin, or of the redemption of the world? Surely such notions are just about as unscientific as they come—are they not?

## Fighting the Fall

According to the Bible, Adam and Eve ate an apple and then everything went pear-shaped. Their decision to reject God's love, authority, and wisdom was a disastrous one for humanity; it resulted in violence, pain, suffering, and loss. Humankind had fallen—fallen out of relationship with its creator, and fallen out of harmony with his creation.

Jesus, as we have seen, stepped in to rescue his people from their self-made mess—and, once they were forgiven and restored, called them to live better lives than they previously had been living. St. Paul took up this line of thought, and instructed those in the earliest churches to "walk in the way of love" and "live as children of light," while "always giving thanks to God" for their salvation (Eph 5:2, 8, 20). The Church had been given a mission: they were to fight back against the darkness that had been brought about by their own rebellion; they were to declare war against the spiritual, moral, and physical effects of the Fall.

It was precisely this mission that was on Francis Bacon's mind as he imagined his New Atlantis. He dreamed of a community committed to bringing back Eden and everything good that came with it—including, of course, our harmonious relationship with nature. Adam and Eve, he believed, had once lived in unspoiled unity with their surroundings; they had owned perfect knowledge of God's creation. After the Fall, they had lost it. But, said Bacon, God's great gift of science could help their offspring to win it back:

> Man by the fall fell at the same time from his state of innocency and from
> his dominion over creation. Both of these losses however can even in this

life be in some part repaired; the former by religion and faith, the latter by arts and sciences.[28]

Bacon was not the only one to think this—in fact, his view was ubiquitous. Peter Harrison summarizes what he found to be the combined mindset of Bacon and his scientific contemporaries:

In order to recover the knowledge of nature lost as a consequence of the Fall and to regain our dominion over natural things, we [i.e., Bacon et al.] presently need special measures to overcome our fallen conditions. These are provided by the kind of experimental science practised by the Royal Society.[29]

There was, however, a problem. The Christian dogma that the likes of Bacon and Hooke assented to included the belief that our rational minds had fallen along with our moral selves—and so we could not, therefore, trust our intuition alone. Logic and reason were still useful, of course, but they were marred; they offered no guarantee of the truth on their own. The Fall had impaired us—we could be mistaken in our deductions.

The solution to this problem—as Harrison hinted at—was "experimental science." A physical trial or demonstration could act as a secondary source of information, or as a safety net underneath a theory, or simply as pleasing confirmation that one's scientific suspicions were indeed correct after all. American–Israeli philosopher of science Noah Efron highlights the power of this seemingly unlikely connection:

Augustine's notion of original sin (which held that Adam's Fall left humans implacably damaged) was embraced by advocates of "experimental natural philosophy." As they saw it, fallen humans lacked the grace to understand the workings of the world through cogitation alone, requiring in their disgraced state painstaking experiment and observation. . . . In this way, Christian doctrine lent urgency to experiment.[30]

From the religious dogma of Adam's and Eve's fallen minds, then, we get the need for practical science sessions. Who would have thought it? Perhaps Genesis chapter 3—forbidden fruit, talking snakes, cursed ground, and all—should be written out, in full, over laboratory doors worldwide.

## Be in Some Part Repaired

The original-sin-led-to-experimental-practice connection is not the only unexpected leg-up that Christian doctrine gave to modern science. The doctrine of God's omnipresence, for example, meant that Newton was prepared to allow his gravity to act across the enormous distances of space, and without obvious bodily contact—a possibility that, up until that point, had been considered highly philosophically suspect. Since God was in some sense everywhere, Newton reasoned, He could bring these things about, and even use natural causes to do so. The great physicist's theology, therefore, was crucial to his model—without it, it is not at all clear that he would have persisted.[31]

The doctrine of the Trinity, too, has made scientists think deeply and then act upon those thoughts. John Polkinghorne, the aforementioned quantum physicist, wrote that "there are aspects of our scientific understanding of the universe that become more deeply intelligible to us if they are viewed in a Trinitarian perspective." Nature's logical ordering, he says, comes from the word of God (Jesus), while its vibrant life hails from the Holy Spirit—and both of these, he suggests, are grounded in the constancy and the power of the Father.[32]

Nearly four centuries before Polkinghorne, Kepler also drew on the inherent notions of unity and diversity among the Father, Son, and Spirit—and went looking for the same sort of interplay in the heavens. Historians of science Peter Barker and Bernard Goldstein write that "theology plays a central role in Kepler's scientific thinking" and that "Kepler's first book cannot be understood without acknowledging its religious dimensions . . . similar issues underlie Kepler's demonstration that the orbit of the planet Mars is an ellipse."[33]

As much as *Conflict* and *Warfare* might protest otherwise, then, there is simply no escaping the truth: Christian dogma has actually played a major part—indeed, many have even argued *the* major part[34]—in establishing the foundations of the science that is so successful today. Draper and White, in setting their literary attack dogs on doctrine, made a terrible mistake—and, as a result, they have done both religion and science a significant disservice.

And now, of course, the damage has been done. We have reached the point where a Nobel laureate feels totally at ease writing an editorial in a national newspaper about why his colleague, purely by virtue of being a Christian, is unfit to work in the sciences. We have reached a point where the Royal Society—which owes its existence to biblical teachings—was so unwilling to tolerate Christianity that Reiss, the target of Kroto's piece, did indeed have to stand down.

How did it come to this? How did *Conflict* and *Warfare*, with their horribly misjudged intentions and their woefully inaccurate storytelling, come to dominate the world of science and ideas? What happened to launch their conflict thesis into the stratosphere when its flaws were so serious and so obvious? And what—if anything at all—can be done about it now?

Perhaps we should draw on a stiff dose of the optimism that inspired Francis Bacon and others all those years ago; perhaps—with the help, of course, of a sufficiently open-minded and willing scientific community—the devastation wrought by Draper and White really can "even in this life be in some part repaired."

# 9

# Agloe, and How to Get Rid of It

I Helped Make It Real • A Medium of Speech That I Can Control • The Father of the History of Science • Light Entertainment at Best • He Too Makes No Historical Case • Rejected Outside the Door • Legitimate Members of the Scientific Community • Whose Fault Is It Anyway? • A Sense of the Real Difficulty • Amen! • Adding Something to the Party • Forgotten Figureheads • The Reconcilers • Goodbye, Agloe

Nearly a hundred years ago, Otto G. Lindberg and his assistant Ernie Alpers rearranged their initials to form an entirely new word—and, just like that, Agloe came into the world.

The duo then wrote their concocted name onto a map they were making of New York State—placing it, quite deliberately, at a rural intersection where nothing of any note whatsoever could be found. Having done this, they published their beautiful, highly accurate, and soon-to-be widely distributed map—a map which showed Agloe as a real town when, in reality, it wasn't.

Why?

Well, because this was the 1930s—and real, first-time mapping on the ground was a rather slow and expensive business. For those who wanted to cut those time-and-money corners, there was a cheaper and easier option: let a rival do all the hard work, and then copy their product—ensuring, of course, that there were enough minor stylistic changes to claim originality.

Agloe, then, was Lindberg's protection: if it ever turned up on someone else's design, he had proof that they had stolen his intellectual property. Sure enough, a couple of decades later, Rand McNally tried to sell such a document, and so Lindberg's company threatened to sue. Rand McNally, though, came up with an unlikely defense: that Agloe, despite what anyone might claim, was *real*.

Remarkably, they were right—for some bright spark had seen the town on the earlier map, decided it might need a store, and set one up at the intersection. Unsurprisingly, given its non-setting, this shop had gone bust—but, because it had called itself Agloe General Store, the lawyers deemed that the town officially existed, even though it didn't.

And so, as bizarre as it all might seem, Agloe has persisted on New York's maps ever since. What's more, the novelist John Green used it as a key location in his smash hit *Paper Towns*, which was later made into a movie—and Agloe, despite not actually being there, has now become a tourist destination. Krystie Lee Yandoli, a writer for the popular website Buzzfeed, journaled in 2015 that:

> Agloe is an important reminder that we get to decide what's important, what exists, and what takes up space in our world. There's so much power in that, and it was a power I felt on that October afternoon. By standing around, asking someone to take my picture in front of the general store sign [the store itself is long gone], observing my surroundings, and stopping to eat a peanut butter and jelly sandwich on a tree trunk across the street from this paper town landmark, I helped make it real. I contributed to making it a reality.[1]

This is, to put it mildly, a rather odd story. Agloe was invented by two long-forgotten men, for an ultimately failed purpose, several generations ago. Their invention inspired others to act upon it—some deviously, some unwittingly, some creatively. One way or another, over the years, these others have conspired to give Agloe an entirely new life of its own—they have "contributed to making it a reality." Agloe may not have been real when it was first drawn up, but now, somehow, it is.

Just like the conflict thesis.

## A Medium of Speech That I Can Control

Who, then, are the people that helped make the conflict thesis real? Who are the Rand McNallys, or the John Greens, or the Krystie Lee Yandolis of the science-versus-God storyline? How did Draper's and White's wrong-headed ideas and fake news come to such prominence that most folk—even some of those considered to be experts—now think that they are wholly undeniable?

Well, as with all good whodunnits, our picture will gradually become clearer as we examine the key characters involved in the plot. So, let's begin our inquiries by meeting a mightily motivated man on a mission: a certain Edward Livingston Youmans.

Youmans was born in New York in 1821, and his parents hoped that he would one day be a Protestant minister. The youngster, however, had different ideas—as the famed Harvard lawyer Charles M. Haar explains:

Concern with "saving people" manifested itself in Youmans as an urge to popularize science. He often assured his mother that the diffusion of science was the most important work in the world, and that, in his own fashion, he was accomplishing moral good. (His devout mother, of course, was never completely convinced.)[2]

The grown-up Youmans stayed true to his dreams—he remained determined to communicate the remarkable recent advances in the sciences to the masses. A chance encounter with the publisher William Henry Appleton (1814–1899) led to a lifelong partnership between the two, and Appleton eventually decided to let Youmans run his own titles. This led to *Popular Science Monthly* and *The International Scientific Series*—both of which were founded in the early 1870s.[3]

Youmans was delighted. Up until now, he had been concerned that science was being dumbed down or compromised before it was introduced to the public at large—the fiction of Jules Verne, for example, mixed genuine discoveries with fantasy—and the New Yorker was frustrated that the great work of his heroes was often presented inaccurately or incompletely, or as a form of tacky entertainment.

His new publications would be different: they would be written by the scientists themselves, and they would be informative, rigorous, and cutting-edge—he would get real science out to the public. "I have long wanted a medium of speech that I can control," he wrote to a friend, "and now I shall have it."[4]

This friend, as it happens, was X-Club founding member, Herbert Spencer—the staggeringly influential philosopher that we first met in Chapter 2. Youmans immediately recruited Spencer as a writer, along with his clubmates John Tyndall and Thomas Huxley. Tyndall—the deliverer of the infamous Belfast address in which he had called for science to replace theology—wrote the first ever *International Scientific Series* title. Spencer—who had called the traditional understanding of Scripture "absurd"—wrote the fifth installment (and, in doing so, invented the phrase "survival of the fittest"). The whole thing was an unmitigated success, with sales in the United States, the United Kingdom, and plenty of other countries going through the roof. This new format was clearly a winner.

Youmans, for his part, was not merely tolerant of Spencer's and Tyndall's disdain for orthodox Christianity; instead, he was a fan of it. He himself was one of the leaders of the controversial Free Religious Association, which had been founded in 1867 to rid humanity of all "dogmatic traditions." This dynamic is crucial to our tale, for it means that the first truly global popularizer of science was also an active promoter of the conflict thesis. The two, therefore, came as a package—right from the very beginning.

In fact, as far as Youmans was concerned, disseminating the latest scientific discoveries and eliminating the firmest religious doctrines were, essentially, two sides of the same coin—and so he was always on the lookout for outspoken writers who were happy to combine the promotion of science with a decent sideswipe at religion.

Does anyone spring to mind?

Fatefully, in 1874, Youmans asked Draper to write the twelfth iteration of the now highly successful *International Scientific Series*. Draper, naturally, agreed—and promptly penned *Conflict*. Then, in 1885, White started writing for Youmans too—indeed, he did so for a decade, producing a series of articles in the *Popular Science Monthly* under the banner "New Chapters in the

Warfare of Science." In 1896, these were edited and compiled into a single volume: *Warfare*.

Our two books, then—the two beating hearts of the conflict thesis—were commissioned, published, and sold by Edward Livingston Youmans, a man who was wholly committed to the abandonment of traditional religion and the elevation of naturalistic science. He set *Conflict* and *Warfare* alongside the work of some of the greatest scientists of the age, and delivered them straight into the households of thousands and thousands of interested layfolk across the planet. Draper's *International Scientific Series* contribution, for instance, went through fifty printings in the US and twenty-four in England; it was translated into ten languages, including French, German, Italian, Dutch, Spanish, Polish, Japanese, Russian, and Portuguese. The God-versus-science narrative—despite all its many faults—was going big.

Still, it's one thing to win over the layfolk—it's quite another to persuade the professionals. Surely the conflict thesis would fizzle out as soon as those in the know got involved—wouldn't it?

## The Father of the History of Science

Before the turn of the twentieth century, there was not really a formal discipline known as the "history of science"—folk had written about it, of course, but there were no university departments of that name, and no one would have called themselves a "historian of science." All that was soon to change, however; and it was to be thanks, almost entirely, to a single individual—George Alfred Leon Sarton.

Belgian-born Sarton, who lived from 1884 to 1956, achieved an extraordinary amount during his academic life. He wrote the mammoth foundational text for his field, *An Introduction to the History of Science*, taking twenty-one years (1927–1948) and five volumes to do so. He founded and was then editor-in-chief of two groundbreaking journals on the subject, *Isis* and *Osiris*, both of which thrive to this day. He established its top professional body, the History of Science Society, in 1924. It's quite the CV.

To this day, the History of Science Society's biggest prize (the winners of which read like a who's who of the very, very, best) is the George Sarton Medal—first presented to Sarton himself by his colleagues in 1955. The medal rewards someone who has made a lifetime contribution in the area, and Sarton was indeed more than deserving of it. He had served innovatively and faithfully, for more than forty years, at Harvard; he put his department and beloved subject squarely onto the world's academic map. Rarely, if ever, has one discipline owed so much to one person. Here, for example, in the very first article in the very first issue of *Isis*, Sarton proclaimed:

> While numberless books, many of them excellent, are published every year on the history of literature, of art, of religions, how is it that there is not yet a single history of science that can be compared with the best of them? When so many institutions, libraries, lectureships have been dedicated to the history of everything, how is it that the history of science has been so much neglected?[5]

It is, to put it simply, impossible to separate George Sarton from the history of the history of science.

What the Belgian thought, therefore—and especially what he wrote and said—really mattered. He was the man setting the terms of the debate; it was he who drew up the history of science landscape. And, crucially for us, he didn't think all that much of religion.

To be more exact, he didn't think much of traditional dogma; nor of the supernatural. Instead, he was a big fan of positivism—the post-revolution, post-Christian worldview of Auguste Comte (see Chapter 2). Although Sarton was not in total agreement with the Frenchman, he did approve of Comte's assertion that we, as a species, need to worship—and that we should, in the absence of God, worship ourselves. Interestingly, the Harvard man even preached as much, at a Church of Humanity in London.[6]

Sarton called Comte the true "founder of the history of science," and broadly agreed with his overall thesis: that men and women, as they develop intellectually, will move away from superstitious nonsense and head instead toward scientific truth (the positive stage).

But he didn't stop there.

If we return to that first ever *Isis* article, we find this:

> The interactions between science and religion have often had an aggressive character. There has been most of the time a real warfare. But as a matter of fact it is not a warfare between science and religion—there can be no warfare between them—but between science and theology ... theologians ... have not ceased from aggravating these misunderstandings.[7]

This sounds rather familiar. And, sure enough, here it comes:

An excellent proof of this has been given in this country. One of the great men of these United States, Andrew Dickson White, has published a splendid book on The Warfare Between Science and Theology.[8]

Yes, that's right—Sarton, the doyen of the history of science, was a committed disciple of White.

He liked Draper, too. In 1952, at the end his career, Sarton sought to pass on the torch to the next band of researchers, and published *A Guide to the History of Science*. This giant tome included a reading list—an elite selection of helpful documents that the great man believed could continue to take the discipline forward. One of his chosen titles was *Conflict*. Explaining the recommendation, he lauded Draper as a "man of science, historian, educator."[9]

And, before we move on from Sarton to look at more recent times, there remains one last point worth noting. For, as it turns out, the launch of his beloved *Isis* had been partially dependent on the support of several prominent scientists who did not believe in God—including Svante Arrhenius, Jacques Loeb, Henri Poincaré, and Wilhelm Ostwald—some of whom were very happy to be outspoken opponents of traditional religious beliefs.[10]

In short, the new academic field of the history of science and the dramatic narrative of the conflict thesis had arrived at Sarton's party hand in hand—and, quite clearly, they planned on sticking around for the slow dances.

By the time, then, that televisions and computers were first getting primed to launch us into the information age, the message of *Conflict* and *Warfare* had taken over. The first true popularizer of science, Edward Youmans, had introduced it to the general populace, and the first true historian of science, George Sarton, had introduced it to the academy. The conflict thesis could claim to be the default position for pretty much everyone. And, since then—thanks to a heady combination of improved technology, greater demand for entertainment, the celebrity scientist phenomenon, rushed-out or poorly researched textbooks, unchecked social media, and increasingly partisan mentalities—it has hardly looked back.

As we shall now see.

## Light Entertainment at Best

As recently as the summer of 2018, the conflict thesis held top spot on the *New York Times* bestseller list. Dan Brown's sci-fi mystery *Origin*—think

*Conflict* and *Warfare* plus sentient computers, Catholic assassins, helicopter chases, and not-so-sizzling romance—took the God-versus-science main theme and turned the dial up to eleven. The initial print run was a remarkable two million copies; the first edition was released in over forty different languages. At regular intervals, the reader is reminded by Brown's enlightened (and extraordinarily attractive) heroes about which side they are supposed to be on:

> The Church's systematic murder, imprisonment, and denunciation of some of history's brightest minds delayed human progress by at least a century.[11]
>
> These conflicts I have described—those in which religious superstition has trumped reason—are merely skirmishes in an ongoing war.[12]
>
> For centuries, most of the devout had looked past vast amounts of scientific data and rational logic in defense of their faith.[13]

*Origin*, in the clear light of day, is really just a good-guys-against-bad-guys fairy tale—but it is one in which the good guys tend to be highly rational, evidence-chasing science-lovers, and the bad guys tend to be dogmatic, evidence-ignoring religious folk. The public, judging by sales at least, have lapped it up—just as they did with its similarly themed predecessors. Brown, it would seem, has struck on a winning formula.

Without Draper and White, or Youmans, or Sarton, however, *Origin* might never have come into being. While its over-the-top action sequences might require some significant suspension of disbelief, the ever-present reason-versus-dogma subtext is much easier to swallow. Why? Because Brown is simply making use of an idea that most of his readers—thanks to his predecessors—probably already "know."

We should keep in mind, of course, that *Origin* and its stablemates are only light entertainment at best. Yes, it would be nice if they did not propagate God-versus-science mythology or breathe yet more life into old lies, but they are fiction first and foremost—so they can, perhaps, be forgiven for frivolously fomenting falsehoods.

What really should bother us, though, is that there is no carefully corrective instruction coming from the world's top non-fiction writers or public scientists about the conflict thesis—there is no calming of the sheep like we find there is with arguments about vaccines, or climate change, or 5G communication towers. Instead, the world's most famous crossover educators— the ones who could, perhaps, help to protect the minds of the general populace for the greater good of us all—often manage to stir up even more trouble and division than there was to begin with. How so?

## He Too Makes No Historical Case

The New Atheist movement—the key members of which are usually named as Richard Dawkins, Daniel Dennett, Sam Harris, and Christopher Hitchens—have, since the turn of the millennium or thereabouts, taken great pride in their assumed roles as instructors of the general public. These thinkers dream of a future dictated by reason alone—if only, they claim, we could overrule our irrational tendencies, then we would enjoy much better lives. They may well be right to think this, of course—but, given the general state of humanity both now and in the past, one suspects that we might never find out.

So, as self-appointed bastions of the truth, have the New Atheists worked hard to set the record straight on science and religion? Have they sought to point out that Draper and White inadvertently led us down the wrong path when it comes to relations between the two? Have they explained the key role that dogma played in developing science? Have they tried to settle matters down, redress the uneven balance, bring us all back together again, and usher in the religious and rational harmony that was longed for by the likes of Bacon, and Newton, and Boyle, and Descartes?

These are precisely the questions that historian of science Ronald L. Numbers and cellular biologist Jeff Hardin decided to drill down on for a study published in 2018.[14] Their findings make rather depressing reading. Dawkins, they discovered, almost entirely ignores the historical relationship between science and religion, focusing instead on his own personal feelings on why faith is always nonsense, no matter who happens to think what, or why, or how.

Dennett comes out only slightly better—he does discuss some sort of "clash" between reason and religion, but Numbers and Hardin complain that he, like Dawkins, is "lacking a historical dimension." Harris, in an apparent attempt to look back a bit further than these other two, confidently asserts that the "maintenance of religious dogma always comes at the expense of science"—and yet, as our authors point out, "he too makes no historical case." They consider the best of the so-called four horsemen to be Hitchens; but even he makes his claims without documentation, they say, and gets basic history wrong.

In a quest to find some New Atheists actually prepared to offer evidence for their view—the claim that "religious dogma always comes at the expense of science"—Numbers and Hardin expand their survey beyond the initial foursome. After working their way through questionable treatises from biologists Jerry Coyne and P. Z. Myers, they arrive at physicist Victor Stenger—who wrote in yet another *New York Times* bestseller that "The totality of evidence indicates that, on the whole, over the millennia the Christian religion was more of a hindrance than a help to the development of science."[15]

This, as we now know, is simply not true. And so, growing increasingly frustrated that not one of the communicators they have analyzed seems to have done any proper historical research whatsoever, our investigators throw the net still wider—hoping, perhaps, to catch some better-informed fish from the right sort of disciplines. They are out of luck: "It might be asked, given the ahistorical accounts of science and religion found in most New Atheist books, whether any historians of science count themselves as members of the group. The answer seems to be no."[16]

Interesting.

Perhaps Peter Harrison, former Andreas Idreos Professor of Science and Religion at Oxford University—who really has done the hard yards—was onto something when he wrote that "The myth of a perennial conflict between science and religion is one to which no historian of science would subscribe."[17]

And perhaps Alister McGrath, the current Andreas Idreos Professor of Science and Religion at Oxford University, was onto something when he wrote that "The idea that the natural sciences and religion have been permanently at war with each other is now no longer taken seriously by any historian of science."[18]

And yet here, of course, we hit a great irony: while a "perennial conflict between science and religion" may not ever have existed in the past—regardless of how much Draper, or White, or Youmans, or Sarton, or Brown, or Harris, or Coyne, or Stenger, or any others might imply or insist that it did—there is, most certainly, a conflict between the two now.

## Rejected Outside the Door

On January 5, 2016, a paper was published in the high-profile internet science journal *PLOS ONE*. Entitled "Biomechanical Characteristics of Hand Coordination in Grasping Activities of Daily Living," it was clearly written by authors for whom English was at best a second language. Here is a section from the abstract:

Drawing a clear functional link between biomechanical architecture and hand coordination is challenging. . . . To explore this link, we first inspected the characteristics of hand coordination during daily tasks through a statistical analysis of the kinematic data. . . . Then, the functional link between biomechanical architecture and hand coordination was drawn.[19]

Two months later, it all kicked off. Danilo Russo, an editor at the journal, posted the following: "I feel my scientific reputation to be put at risk by this incredible mistake, so should this paper not be retracted as soon as possible I will be compelled to resign."[20]

Russo was not the only one outraged. James McInerney, a prominent microbiologist, tweeted that *PLOS ONE* "is now a joke." Enrico Petretto, a geneticist, agreed: "if the paper isn't retracted, my students, collaborators and I will have no choice but to refrain from considering (i.e., reading, reviewing and citing) papers published in *PLOS ONE*."[21]

Sure enough, the journal took swift action. The anonymous academic editor who had approved the paper was tracked down and promptly removed from his or her post.[22] The paper itself was pulled—it now has a big red box sitting over the top of it marked "Retraction." The explanation given includes the following: "The *PLOS ONE* editors consider that the work cannot be relied upon and retract this publication. The editors apologize to readers for the inappropriate language in the article."[23]

Embarrassed and upset, one of the paper's authors felt the need to make peace with the scientific community. Mingjin Liu, a Chinese biomechanical engineer, begged for forgiveness and for acceptance:

Your pressure to PLOS ONE to retract my paper, which is only due to the misunderstanding of the word, has hurt deeply my academic career. I am just a young scientist. I hope to have a chance to correct my mistake, not to be rejected outside the door.[24]

What, then, was "the word" that caused such a storm? To find out, we can head back to that awkwardly written abstract:

The explicit functional link indicates that the biomechanical characteristic of tendinous connective architecture between muscles and articulations is the proper design by the Creator to perform a multitude of daily tasks in a comfortable way.[25]

And there it is: "Creator." The same "Creator" that Isaac Newton wrote about in *Principia*, that James Clerk Maxwell credited his electromagnetic success to, and that Rosalind Picard relies upon for her computer science today. The problem? Well, this was a scientific paper in a scientific journal—and, as everyone post-Youmans and post-Sarton knows, science and religion are at war.

So strong was the feeling—a whole battalion of scientists tweeted their anger en masse[26]—that it didn't even matter when the authors all came out and explained that they weren't creationists, that they hadn't intended to reference God in the first place, that the word they had in mind was a Chinese one for Nature, and that they would happily have changed the phrasing had they been given the opportunity. It also didn't seem to matter what the actual method, results, or conclusions of the study were, for the retraction notice does not mention any of these as a problem. Even the aforementioned New Atheist thinker P. Z. Myers admitted that "there's nothing wrong with the data that I can see."[27]

Mingjin Liu and his collaborators, we must conclude, were simply collateral damage—they wrote the wrong word at the wrong time in the wrong place, and paid a high price for it. Their story is a rather sad one. The intellectual blast cloud first sent up by the explosive writings of Draper and White more than a century ago is still, to this day, depositing its toxic contents all over the place. And, like most toxic clouds, it is causing a fair bit of misery.

## Legitimate Members of the Scientific Community

Perhaps the best solution to all this is to calm down a little bit, and lock the science and religion debate up in the nearest ivory tower. After all, isn't that where it belongs? Isn't it the domain of professors with tenure taking breaks from their fireside naps to write snooty, port-stained papers at one another? Why should anyone in the real world care in the slightest about John William Draper, or about Andrew Dickson White, or about what the two of them wrote, or about who read their work—or even about anything that has happened because of them since?

Well, because there has been an impact in the real world, that's why—and it hasn't been a good one.

We have already seen how some people have fallen victim to the conflict thesis. Michael Reiss lost his job, as did that nameless *PLOS ONE* editor. Tom

McLeish, according to Jerry Coyne, should not have been given his role at the Royal Society. Mingjin Liu and co. had their paper retracted for no apparent scientific reason. And these, rather depressingly, are hardly isolated cases.

When Francis Collins, the director of the Human Genome Project, was appointed as the new director of the National Institutes of Health by President Barack Obama in 2009, he faced immediate opposition. The *New Yorker* magazine reported that:

> Many of his colleagues in the scientific community believed that he suffered from "dementia." Steven Pinker, a cognitive psychologist at Harvard, questioned the appointment on the ground that Collins was "an advocate of profoundly anti-scientific beliefs." P.Z. Myers . . . complained "I don't want American science to be represented by a clown." Collins's detractors did not question his professional achievements, which long ago secured his place in the first rank of international scientists.[28]

Collins's "dementia" here appears to be another name for his supposedly "anti-scientific" Christian beliefs. These are, we should not forget, the same beliefs that drove Kepler and Descartes and Hooke and Faraday toward their scientific breakthroughs—and the same beliefs that underpin the methodology that Pinker and Myers still use to this day.

If such bullying and name-calling can comfortably exist at the most visible of public levels, one wonders what things might be like down at the coalface. The answer is that they aren't always great. The eminent sociologist Elaine Howard Ecklund analyzed the experiences and views of more than ten thousand scientists around the world—a huge grouping across eight countries, and one which included representatives of almost every conceivable worldview. She found that those with religious beliefs often felt themselves under pressure in the workplace; in her 2019 book, *Science and Secularity*, she writes:

> Discrimination can take many forms, such as denying access to opportunities like jobs or using derogatory and abusive language. . . . Religious scientists in the United States also noted occurrences of "unnecessarily vocal Christian bashing" and stated that religious scientists "have to have a thick skin in order to be in the science realm." . . . A female graduate student in physics explained to us that being religious "opens you up to be not taken seriously."[29]

Ecklund is distinctly unimpressed by this, and doesn't pull her punches: "Nonreligious scientists use mockery to classify themselves as legitimate members of the scientific community while religious scientists are not."[30]

The largest religious grouping in Ecklund's study were Protestant Christians—and 40% of them (in the United States) claimed to experience some form of discrimination while carrying out their scientific roles. Given the huge part that Protestant theology played in the development of modern science, this is an extraordinary cultural change. The idea that an orthodox believer might one day be overlooked for a promotion, or called to resign, or have papers pulled, or be subjected to open hostility would have been wholly unimaginable to many previous generations of scientific greats. Here we are, though, and it is really happening.

But what if the root of the problem lies not with bullying biologists, or with caustic chemists?

What if it actually lies with the Christians themselves?

## Whose Fault Is It Anyway?

When Jerry Coyne was invited to give a speech at the Freedom from Religion Foundation's annual convention in 2016, he issued a heartfelt cry:

> You write a book on evolution with the indubitable facts showing that it has to be true, as true as the existence of gravity or neutrons, and then you realize that half of America is not going to buy it no matter what you say. Their minds cannot be changed. . . . So you start realizing that religion is perverting what you're trying to do with science by making statements about the world, but then supporting them with various cockamamie methods. And so you become an atheist.[31]

Coyne, like many others, despairs at the agenda he sees and feels around him—these Christians, he says, are constantly attacking science, constantly undermining science, and holding everyone back with their divinely inspired nonsense. Even those who claim to be pro-science aren't, he laments: "They say intelligent design, but what they really mean is Jesus."[32]

It is hard not to sympathize with Coyne here—he has committed his life to the careful study of data and evidence; he has tried to share the resulting excitement and joy with the world; he has met with obfuscation, skepticism, and anger; he has found it impossible to change his opponents' minds, even when they are presented with a plethora of "indubitable facts."

Coyne's complaints are hardly rare. In his multi-million-selling *The God Delusion*, Richard Dawkins voices them too: "If this book works as intended, religious readers who open it will be atheists when they put it down. What presumptuous optimism! Of course, dyed-in-the-wool faith-heads are immune to argument."[33]

The real hot potato here, it must be said, is evolution. The evolution-versus-creation debate seems to be where it all (currently, at least) comes to a head; it is where we find the assorted combatants getting the most wound up; it is where we stumble across the nastiest insults; it is where we find much of the conflict thesis ink being spilled—by writers, incidentally, on both sides of the divide.

In America in particular, the polarization can be quite extreme. Arguments about the teaching of evolution and of creation in schools have been played out in the courts on a number of occasions, with the gavel coming down in favor of evolution each time. Despite this, a reasonable proportion of science teachers choose either to avoid mentioning the topic or, sometimes, to actively speak against it. In 2013, *The American Biology Teacher* journal felt the need to remind the nation's staff about the legal requirements of their role:

> Teachers who advocate alternatives to evolution are not teaching within the confines of the law . . . avoiding evolution may be unlawful if teachers appear to be doing so for religious reasons. Even if teachers are avoiding evolution without the perception of doing so for religious beliefs, they can be ordered to do so to ensure that state and national standards are addressed.[34]

This is not the kind of instruction that would show up in a happy, harmonious environment. And, while evolution grabs most of the attention, there seem to be other pinch points too. Here is Dawkins again, this time on his website:

> Vocal anti-vaccine propagandists have already fostered outbreaks in the US and UK, as gullible parents choose to leave their children unprotected. Religious dogmatists have long been among those leading the charge against the advancements of science and medicine, hiding behind the "right" to practice their faith.[35]

And then there's climate change. In 2016 *Worldviews* journal ran an in-depth study of attitudes commonly held by those who assent to traditional doctrines. Here is a typical finding:

> Catholic Answers Forum (CAF) is a popular site for Catholics, with nearly half a million members . . . skeptics start most of the [climate change] threads and write most of the posts. Some of these threads include: "Nature's Wrath? Most Say Cycles, Not Climate Change to Blame" . . . and "New Documentary Shows How Leftists Use Fake Science to Enslave Us via Climate Change Fraud."[36]

The conflict thesis, it would appear, has done an Agloe: it wasn't real in the past, but it has become real today. And perhaps, rather than Draper or White (or Youmans, or Sarton, or Sagan, or Dawkins) being to blame, this development might even be the fault of modern Christendom itself—for it seems to be railing against science.

What can we say, then? If Christian scientists really are missing out on the top jobs, or if they are being laughed at in the workplace, or if they are often treated with rampant suspicion by their atheist or agnostic colleagues, should they honestly be all that surprised? After all, their lived experience—as difficult or upsetting as it might well be—seems to be a mess entirely of their church's own making.

Perhaps Draper, despite all his other errors and inaccuracies, was actually correct in his assessment of rational science and dogmatic religion: "One must yield to the other; mankind must make its choice—it cannot have both."

So is that it, then? Are we done?

No, not quite.

## A Sense of the Real Difficulty

Clearly, there are plenty of folk out there, from both sides, actively pushing the conflict thesis. What's more, they seem to bring out the worst in each other—it feels like we are a long, long way from the science-and-religion harmony of Francis Bacon's *New Atlantis*. Perhaps, though, we can still end up there; perhaps all it would take is a tweak of the rudder.

When Coyne was giving his speech to the Freedom from Religion Foundation, he made the following declaration: "The fact of evolution is not only inherently atheistic, it is inherently anti-theistic. It goes against the notion that there is a God."[37]

This, by the way, is false—evolution, being a strictly scientific theory, is unable to comment on what its own ultimate driving force is. In the same way that Kepler considered God to be the reason behind the orbits he derived, many current biologists—including, incidentally, Francis Collins—believe that God is behind evolution. Indeed, the evolutionary model is the favored position of the Catholic Church today; it has been since the 1950s.[38] And the Catholic Church, despite what Coyne might say, tends to accept "the notion that there is a God."

Still, his overall message has clearly got through, as M. Elizabeth Barnes has recently demonstrated. Barnes is well placed to comment: she is a biologist, a historian of science, and a philosopher of science. Indeed, her specialism just happens to be the various attitudes that exist toward evolution within education. In 2019, she and her team of co-authors found that more than half of the two thousand or so college students they asked believed that evolution was necessarily atheistic. One of the group's conclusions is particularly enlightening: "Highly religious students who thought evolution is atheistic were less accepting of evolution by all measures compared with highly religious students who thought evolution is agnostic."[39]

Well, this is interesting—it seems that Coyne and co. might actually be shooting themselves in the foot. Maybe, if they really want religious people to accept evolution, they should stop insisting that evolution is irreligious. If anything, this seems to be a rather obvious point—and it is not the first time that it has been made. Here is theologian Thomas McCall:

> It is a widespread truism that "religion and science are in conflict," and, for many people, evolution and traditional Christian belief are locked in a death match. Accordingly, given these options, many traditional Christians reject evolution as a viable option.[40]

So why, given this observation, doesn't Coyne change his pitch? Is it, perhaps, that he doesn't really want to—and that he is every bit as stubborn and inflexible as the "half of America" that he is frustrated with? One could be forgiven, at this point, for siding with Andrew Dickson White: "Then it was that there was borne in upon me a sense of the real difficulty—the antagonism between the theological and scientific view of the universe."[41]

Antagonism indeed. Each group seems to be pushing the other further and further away—and, as they do so, they are making the prospect of a shared understanding look more and more like a pipe dream.

In fact, right now, instead of a shared understanding, we have papers retracted for accidentally mentioning a creator; we have enforced resignations; we have accusations of "dementia"; we have discrimination in the workplace. We have our intellectual guardians loudly promulgating wrong ideas; we have a majority of school students declaring, incorrectly, that "the scientific view is that God does not exist."[42] We have evolutionary theory hauled out, by both sides, as a shibboleth—and anyone who gets the answer wrong is considered as good as dead to their community. In short, we have conflict. We have warfare.

But does it really have to be like this?

Well, in 2014, Google removed Agloe from their maps—having discovered, presumably, that there was no such town in the first place.

So maybe—just maybe—there is hope.

## Amen!

"The aim of the physicist is to explain all," writes our author. "Although much remains to be done . . . there seems to be no reason to doubt that in principle the scientific picture can be complete."[43]

These are bold words. And there are more:

> Early scientists would invoke God as the explanation of phenomena their science could not explain . . . [since then] the need to account for the unexplained in terms of God has continued to recede. . . . Now everything seems at least in principle amenable to scientific investigation.[44]

"Amen!," as Neil deGrasse Tyson wouldn't say.

The person who wrote these words was the recently deceased Sir John Houghton (1931–2020). A pioneering physicist, Houghton headed up the Intergovernmental Panel on Climate Change from 1988 until 2002, saying of its first report in 1990—which he was the editor of—that "it became very clear to us that there was some real danger ahead."[45]

When the global response didn't kick in as he felt it should, Houghton pushed harder. In 2003, he accused the US government of "an abdication of leadership of epic proportions" and of "refusing to take the problem seriously."[46]

Then, in 2007, with the world beginning to wake up to the problem, he shared the Nobel Peace Prize for his work—and, two years later, he was awarded the Albert Einstein World Award of Science, which "takes into special consideration research which has brought true benefit and wellbeing to mankind."[47] He died—of coronavirus—as a scientific hero; his legacy is a welcome reminder of science's profound value to us all.

Houghton has a soulmate in Katharine Hayhoe—who, like him, is a physicist-turned-climate scientist. Named as one of *Time* magazine's 100 most influential people in 2014, her TED talk on climate change has been watched more than four million times, and she has the ear of the brightest and best all over the planet. On her personal website, she adds her voice to Houghton's by urging the world's leaders—as well as the rest of us—to take action:

> I don't accept global warming on faith: I crunch the data, I analyze the models...
>
> The data tells us the planet is warming; the science is clear that humans are responsible; the impacts we're seeing today are already serious; and our future is in our hands.[48]

However, had Harry Kroto had his way, then Houghton and Hayhoe might never have seen the inside of a laboratory, let alone end up with a big scientific job. For both of them, as it happens, are fully paid-up, Bible-believing, faith-promoting, God-squadders.

## Adding Something to the Party

Hang on a second—isn't this just cheap point-scoring? Why does it matter that two important climate scientists also have personal beliefs in God, and in water turning into wine, and in the resurrection, and in an eternal life to come? What difference does that make to our conflict thesis discussion?

Well, it does seem that there are several points worth dwelling on. Firstly, it is clear that Houghton's and Hayhoe's faith did not hold them back from doing proper science. Secondly, if there is a concern that some Christian groups are too quick to be suspicious about climate change, then surely it is a good thing to have active and committed believers ready to engage with them on some sort of common ground—is it not? And, thirdly, both of them explicitly link their beliefs to their work. Here's Houghton:

> What we're doing as Christians is exploring our relationship with the person who is the creator of the universe. Now that's something that is absolutely

wonderful. . . . I'm very concerned about the people in the much poorer countries of the world who are very disadvantaged by [climate change]. . . . As a Christian I feel we have a responsibility to love our neighbours and that doesn't just mean our neighbours next door.[49]

And now Hayhoe:

Climate change disproportionately affects the poorest and most vulnerable people in the world . . . how can I truly be loving my neighbour if I close my eyes to this issue? The reason I couldn't was because of my faith. So that's why I do what I do, and that's why I am who I am. The reason why I am a climate scientist is because I'm a Christian, not in spite of it.[50]

Do we really want to kick people like this out of science? Surely not. In fact, it is hard not to conclude that Houghton and Hayhoe are adding something to the party—not subtracting from it.

Is there some sort of way, then, that we can have both? Can we embrace the best of religion and the best of science, in unison? Can we fight against alienation and separation, and have unity in its stead?

Can we, after all these generations, finally get rid of Agloe?

Yes. Sure we can.

## Forgotten Figureheads

Before we approach that endgame, however, it is probably worth reminding ourselves about what we have learned so far. We have been on quite the tour through the varied landscapes of the history of science and religion—and we would do well, now, to remember some of the highlights.

Firstly, at the beginning of our adventure, we came across two nineteenth-century gentlemen who had, somehow, managed to fool much of the world—many experts included—right up to the present day. We have gradually come to know John William Draper and Andrew Dickson White as people, and we have also developed a feeling for their two deeply flawed manuscripts: *A History of the Conflict between Religion and Science* and *A History of the Warfare of Science with Theology in Christendom*. Our duo, we discovered, did not work in isolation, for many others were having the same conversations—but, one way or another, their twin totemic texts ended up being the most virulent vectors of the argument.

*Conflict* and *Warfare* spun cautionary tales of flat Earths, banned autopsies, Dark Ages, persecuted scientists, burned books, dangerous dogmas, abandoned classics, excommunicated comets, imprudent popes, and unlikely unicorns—in short, of a never-ending battle between science and religion.

But, bit by bit, we have seen that these tales were either manufactured, or mistold, or misleading, or malicious. Dogma and science have enjoyed a peaceful relationship much of the time–indeed, Christian doctrine helped to provide both the presuppositions and the motivations that got modern science going in earnest.

And, in an unexpected twist, we learned that both Draper and White were actually trying to do religion a favor, rather than bash it. Their projects, both men confidently announced, would eventually reconcile faith to science. In fact, drawing deep on their liberal Protestant capital, each man proposed his own highly compromised, science-friendly, pseudo-Christianity—Draper's governed robotically by his head, and White's guided gently by his heart.

Yet such woolly ideas were doomed from the beginning: a Christian gospel without a divine Christ, or a reliable Bible, or a real resurrection is a gospel drained of both its identity and its power and, as such, is no good to anyone. Plenty of their readers spotted this. Traditional thinkers wrote the books off as unholy nonsense; progressive thinkers skipped over them, like stepping stones, on their way to agnosticism or, more often than not, atheism.

Then, in this final chapter, we have followed what has happened ever since. Thanks largely to a driven publisher of popular science—Youmans—and a driven professor of the history of science—Sarton—the myths and the errors of Draper and White began to take over like an invasive species. During the course of the twentieth century, and then into our own, they were promoted again, and again, and again—so much so, that the once fictional conflict actually managed to turn itself into a reality.

And so now, there are scientists wary of religion. Now, there are believers wary of science. Every time one of the falsehoods from *Conflict* or *Warfare* is carelessly repeated—as frequently happens in both bestselling fiction and non-fiction, often from folk who should know better—this wariness increases yet further.

The results of all this are non-trivial: job losses, bullying, and discrimination; vaccination conspiracies; climate change controversy. Draper and White have won the battle for our minds—but it is an unholy victory that they neither intended nor foresaw. They are now forgotten figureheads ruling anonymously over a divided empire that they never would have wanted.

If the tour ended here, on that rather glum note, we may well feel bereft of all hope. Happily, though, it doesn't—for it is now time to consider the future. Is there, perhaps, a different way forward for us altogether?

In fact, let's rephrase that last question: is there a different way forward for us, *all together*?

## The Reconcilers

We must guard, of course, against blind optimism here. After all, the conflict thesis is everywhere—Draper and White have taken over the world. And yet, if we allow ourselves a closer look, there are indeed some reasons to be positive. Out there, at work in our universities, schools, laboratories, and faith communities, are a whole cohort of anti-Drapers and anti-Whites pulling together a library of anti-*Conflicts* and anti-*Warfares*.

So, who are these reconcilers? And what, exactly, are they up to?

Well, two of them are the aforementioned Francis Collins and his colleague Deborah Haarsma. In 2007, Collins founded BioLogos—an organization committed to healing some of the damage meted out by the prevalence of the conflict thesis. BioLogos communicates cutting-edge science to the Christian community, encouraging them to embrace it rather than fear it. Haarsma, an astrophysicist, is its current president. A few years ago, she spoke at Bethel University—an American Evangelical institution—about her astronomy:

> To me as a Christian, those orderly laws that I fell in love with . . . are a display of God's faithful governance. In Jeremiah, God points to the fixed laws of heaven and earth as a testimony that he will faithfully keep his covenant with Israel (Jer. 33:19–26) . . . when I sit down at the computer to analyze data, in no way am I setting aside my faith. In fact, my faith calls me to study those natural laws as a way of celebrating the faithfulness of God.[51]

Here, then, is a Christian in full-blown evangelism mode, seeking to win back the lost—but, in this particular case, to win them back to science. Haarsma hopes for reconciliation. In fact, BioLogos gives its mission statement as inviting "the church and the world to see the harmony between science and biblical faith."[52]

Similarly, the Scientists in Congregations (SiC) group in the United Kingdom has sought to bring the joy of science to its religious community.[53] Reverend Professor David Wilkinson—astrophysicist and priest—gives a heartening example of their activities:

> A former President of the Baptist Union started to introduce science experiments into his local church's children's work. This was a strong message to children and adults: it affirmed the role that scientists play, and broke down the impression that science is a threat. SiC teaches that science is a vocation.[54]

Lizzie Henderson and Steph Bryant, two biologists working for Cambridge's Faraday Institute, spend time in schools talking about science and faith with young people. Aware of the nervousness or suspicion that can be present, Bryant remarks that she and her colleagues look to create environments in which students "feel safe to voice their biggest, most vulnerable questions and doubts" and are able to "explore what the different human ways of searching for truth—from science, to philosophy, to faith—can bring to the overall big picture."[55]

There are many more initiatives like these rising up, all over the world, and they are not hard to find. Recently, one which has been making a few headlines is Peaceful Science—a community partially overseen by Joshua S. Swamidass, a computational biologist at Washington University. In 2017, Swamidass made waves by publishing new research which allows room for both the current consensus understanding of evolution and also the traditional view of Adam and Eve: a human pairing, without parents, created in the Middle East less than 10,000 years ago—and from whom we all would truly descend.[56]

Hmmm. This sounds rather like dodgy Christian apologetics dressed up in an ill-fitting lab coat—and that is certainly a common occurrence—but Swamidass's breakthrough is different. Prominent atheist biologist and science communicator Nathan H. Lents explains why:

> Swamidass is not peddling pseudoscience. Indeed, earlier this year, he and I teamed up on the pages of Science to rebut claims by evolution critics. [His work] went through a rigorous process of open peer review, involving scholars from many diverse disciplines and even some secular scientists,

including myself and Alan Templeton, a giant in the field of human population genetics. Invited to find fault in his analysis, we couldn't.[57]

Of course, it is very early in the game for us to predict where Swamidass's
research will end up taking everyone—but what is so interesting about this
is that he and Lents have become friends, and that they are treating this as an
opportunity to bring about togetherness in their shared world. Indeed, Lents
is on the advisory board for Peaceful Science, which says it aspires toward
"humility, tolerance, and patience."[58] Swamidass himself says:

> Students, pastors, and scientists contact me to explain how much my work
> has helped them. It relieves a major point of tension that many have strug-
> gled with for a long time. What's more, it leaves us with a grand opportunity
> to explore how sacred and natural history can be understood together. In
> place of the conflict, this is the hopeful conversation that is reinvigorated
> now, with invitation to us all.[59]

It is not just today's scientists, by the way, who are directly taking on the
conflict thesis—for the historians of science are continuing, almost daily, to
pick it into ever smaller pieces. Mark A. Noll and David N. Livingstone—
both of whom were part of the first post-Sarton wave—wrote in 2018 that
there are now "numberless scholars who seek a better way."[60]

Within a generation, it will be this new breed of academics who are leading
history of science departments the world over. Having had their eyes opened by
the likes of Margaret Osler, Ronald Numbers, Peter Harrison, and more, they
will be able to teach their own students about how Sagan and Coyne and the
New Atheists had merely been led down Draper's and White's garden path. Such
a change is highly significant—and, hopefully, it will begin to percolate down
through the system. Perhaps textbooks, at least, will begin to sort out their acts.

One of those who seek this "better way" is Londoner Nathan "Bassicks"
Bossoh. A former music student—hence his nickname—Bossoh has
switched disciplines to the history and philosophy of science, and found it to
be a very happy home. The current object of his research is George Campbell
(1823–1900), the eighth duke of Argyll, and a contemporary of Draper and
White. Argyll made progress in aeronautics, ornithology, evolutionary biology, anthropology, geology, geography, theology, and education—all while
actively pursuing his Christian faith. Reflecting on this, Bossoh sees reasons
to be positive about the future:

There is certainly no inherent reason as to why the Conflict Thesis cannot be overturned. The literature has already largely dismantled it, but now younger scholars are increasingly working to push this beyond the academic sphere by utilizing the new tools of the internet age. Thomas Huxley trained up his succeeding generation to perceive conflict between science and theology; today we can train our succeeding generations to understand that the relationship is far more positive than that.[61]

The "tools of the internet age" that Bossoh describes include blogs—such as Tim O'Neill's *History for Atheists* site. O'Neill is an atheist himself, and had become so fed up with seeing fellow non-believers consistently churn out Draper-and-Whiteisms that he started setting the record straight online. O'Neill is both amusing and acerbic in his analysis, and has won the respect of many for his no-nonsense approach. Cambridge classicist Tim Whitmarsh puts it this way:

> Getting history right is crucial, and no one—neither the religious nor the irreligious—should get a free ride when it comes to instrumentalising the past. Tim O'Neill's forthright blog does a valuable job in keeping us all honest.[62]

And, on the topic of keeping us all honest, Ecklund's data turns out to be extremely useful. For, across all of her thousands of scientists, and across all eight countries, only a quarter of the respondents said they believed there was conflict between science and religion. By contrast, a whopping 68% felt that there was not.[63] The professionals themselves, it would seem, do not agree with Coyne, or Dawkins, or Kroto.

Perhaps, then, there are indeed reasons to be hopeful. The conflict thesis might not stick around forever—not if these reconcilers have anything to do with it. The likes of Haarsma, Swamidass, Lents, Bryant, Bossoh, O'Neill, and Ecklund are actively challenging it on a daily basis—and they are being joined by new and imaginative allies all the time.

But can they win? Can they actually unfool the world?

## Goodbye, Agloe

Well, it will be tough. The big problem—in fact, the biggest problem—with undoing the work of Draper and White is the sheer scale of the task. And, by scale, we are really talking exposure. The combined audience of the conflict thesis acolytes utterly dwarfs that of the reconcilers mentioned above—and the disparity is really quite extraordinary.

The New Atheists' collusion of texts, for instance, outsell the BioLogos team's efforts by several orders of magnitude. Sagan's *Cosmos* had 500 million viewers in the 1980s; Tim O'Neill gets a few thousand on YouTube today. Dan Brown's sales figures, for title after title, are astronomical—yet how many men or women on the street could even name a novel in which science and religion are friends?

Still, despite the immensity of the task, it is undoubtedly worth the effort. For this is about bringing people together and, ultimately, it is also about the truth. Our world does not need more conflict—especially false and manufactured conflict. It does not need more separation, or more wariness, or more division, or more suspicion. Here is Bossoh again, passionately explaining how the ongoing fight to debunk Draper and White is about more than just academic squabbles or being technically right:

> Having grown up in an African church, I have seen that one of the problems with the Conflict Thesis is that it has implicitly (and explicitly) hindered black Christians from engaging with science in various ways—because there is an inherent notion that science is against their faith. This was certainly the case for me for a long time. Dismantling the Conflict Thesis would, I think, help black Churches (and individuals) to engage more positively with science in relation to their beliefs.[64]

Theologian Rebecca McLaughlin concurs. After poring over Ecklund's vast sea of data, she applies the findings to her own field: "The narrative that presents science as antithetical to Christianity is part of what is keeping underrepresented groups (African Americans, Latino Americans, and women) out of the sciences."[65]

Well, then. Here we are, with a known lie that is now well into its second century and still going strong. This lie is the cause of open discrimination in the workplace. It has misled generation after generation of school students, and continues to push potentially gifted scientists away from the subject before they have time to discover a love for it. It separates and divides; it upsets and worries.

And yet the corrected version of Draper and White, by contrast, includes and inspires. We have Church Fathers performing controlled, long-term experiments on peacock meat all the way back in AD 400. We have bishops writing river-crossing puzzles more than a thousand years ago. We have medieval monks quietly committed to keeping ancient Greek brilliance alive.

We have clergymen refusing to limit God, and opening the world's eyes to an infinite set of possibilities in the process. We have grand and utopian visions of scientific islands in the 1600s—all driven by a commitment to the Bible. We have leading logicians inventing invaluable new ways of thinking rationally—and promptly using these novel techniques to write down proofs of the goodness of God.

Even in its own right—even if there was nothing else to be gained other than the theater of it all—it is already a story worth telling. Yet there *is* more to be gained. For, where the conflict thesis divides us, the truth can draw us together. Where the conflict thesis worries us, the truth can bring us peace. Where the conflict thesis restricts us, the truth can set us free.

So, isn't it time we took Agloe off all of our maps?

After all, it isn't real.

# Notes

## Chapter 1

1. Bill Kaysing, *We Never Went to the Moon: America's Thirty Billon Dollar Swindle*. (An edition of this title has been recently republished by CreateSpace.)
2. *Conspiracy Theory: Did We Land on the Moon?* Fox TV, Nash Entertainment, February 15, 2001.
3. See space.com, for instance: https://www.space.com/12796-photos-apollo-moon-landing-sites-lro.html
4. David Robert Grimes, "On the Viability of Conspiratorial Beliefs," *PLoS ONE* 11, no. 1 (2016): e0147905, https://doi.org/10.1371/journal.pone.0147905
5. *Attitudes Toward Space Exploration—Ipsos Poll on Behalf of C-SPAN—Belief in Authenticity of the 1969 Moon Landing*, C-Span, July 10, 2019.
6. Sheldon Amos, "The Intellectual Development of Europe," *Westminster Review* 27, no. 1 (January 1865): 94–142.
7. Amos, "The Intellectual Development of Europe."
8. John William Draper, *History of the Conflict between Religion and Science* (New York: D. Appleton and Co., 1874), vi, xi, 363 (emphasis added).
9. *Report of the Committee on Organization, Presented to the Trustees of the Cornell University. October 21st, 1866* (Albany, NY: C. Van Benthuysen & Sons, 1867), 48.
10. Andrew D. White to Ezra Cornell, August 3, 1869, Andrew Dickson White Papers, Division of Rare and Manuscript Collections at Cornell University, reel 9 (hereafter cited as White Collection, and reel number).
11. "First of the Course of Scientific Lectures-Prof. White on 'The Battlefields of Science,'" *New York Daily Tribune*, December 18, 1869, p. 4
12. Andrew Dickson White, *A History of the Warfare of Science with Theology in Christendom*, 2 vols. (New York: D. Appleton and Co., 1896), 1.v.iii.
13. Ronald L. Numbers, ed., *Galileo Goes to Jail and Other Myths about Science and Religion* (Cambridge, MA: Harvard University Press, 2009), 1, 6.
14. Keith S. Taber, et al., "To What Extent Do Pupils Perceive Science to Be Inconsistent with Religious Faith? An Exploratory Survey of 13–14 Year-Old English Pupils," *Science Education International* 22, no. 2 (2011): 99–118.
15. Richard Dawkins, *Science in the Soul: Selected Writings of a Passionate Rationalist* (New York: Random House, 2017), 277.
16. Dan Brown, *Angels and Demons* (London: Corgi, 2001), 43.
17. Peter Byrne, *The Many Worlds of Hugh Everett III* (Oxford: Oxford University Press, 2010), 385.

18. Allan Chapman, *Slaying the Dragons: Destroying Myths in the History of Science and Faith* (London: Lion, 2013), 100–102; Owen Gingerich, *God's Planet* (Cambridge, MA: Harvard University Press, 2014), 7–56.

19. David Hutchings, "Storytelling and Galileo," in *A Teacher's Guide to Science and Religion in the Classroom*, ed. Berry Billingsley, Manzoorul Abedin, and Keith Chappell. London and New York: Routledge, 2018, 63–65.

20. Chapman, *Slaying the Dragons*, 103–105.

21. Samuel Klumpenhouwer, "Early Catholic Responses to Darwin's Theory of Evolution," *Saeculum* 7, no. 1 (December 2011). www.saeculumjournal.com/index. php/saeculum/article/view/11311/13001

22. David Aaronovitch, *Voodoo Histories* (London: Vintage, 2009), 15.

23. Andy Coghlan and Priya Shetty, "Royal Society Fellows Turn on Director over Creationism," *New Scientist*, September 16, 2008.

24. Elaine Howard Ecklund et al., "Religion Among Scientists in International Context: A New Study of Scientists in Eight Regions," *Socius: Sociological Research for a Dynamic World* 2 (September 2016): 1–9.

25. Elaine Howard Ecklund et al., *Secularity and Science: What Scientists Around the World Really Think About Religion* (New York: Oxford University Press, 2019), esp. 46–47 as an example of the discussion.

26. https://whyevolutionistrue.wordpress.com/2017/01/21/yet-another-accommodationist-book/

27. White, *A History of the Warfare*, 1.xii.

28. Draper, *History of the Conflict*, 364.

# Chapter 2

1. This almost certainly did not happen—but Hartlepool remains famous for it nonetheless. Residents are still referred to as monkey-hangers.

2. Isaac Newton and Robert Hooke, Isaac Newton Letter to Robert Hooke, February 5, 1675 (Newton and Hooke disliked each other, intensely so at times. It is quite possible that Newton's famous statement is not one of humility at all, but was intended to poke fun at Hooke—who was very short). See, for instance, the discussion in Alexandre Koyré, "An Unpublished Letter of Robert Hooke to Isaac Newton," *Isis* 43, no. 4 (December 1952): 312–37.

3. Report No: 8/1988. *Report on the accident to Boeing 737-236, G-BGJL, at Manchester Airport on 22 August 1985*, 2 (see: https://assets.publishing.service.gov.uk/media/5422efe840f0b61342000277/8-1988_G-BGJL.pdf).

4. *Searching for Starlite*, BBC Television: bbc.com/reel/playlist/searching-for-starlite.

5. *Tomorrow's World*, BBC Television, March 8, 1990.

6. *Searching for Starlite*.

7. *Searching for Starlite*.

8. Jeff Hardin, Ronald L. Numbers, and Ronald A. Binzley, *The Warfare Between Science & Religion* (Baltimore: Johns Hopkins University Press, 2018), 6.

9. Hardin, Numbers, and Binzley, *The Warfare Between Science & Religion*, 123.

10. Hardin, Numbers, and Binzley, *The Warfare Between Science & Religion*, 6.

11. Hardin, Numbers, and Binzley, *The Warfare Between Science & Religion*, 143 (Robert's quote is from *Universalist Quarterly and General Review* n.s. 12 (1875): 252).

12. William Doyle, *The Oxford History of the French Revolution* (Oxford: Clarendon Press, 1989), 259.

13. See the discussion in Mona Ozouf, *Festivals and the French Revolution*, trans. Alan Sheridan (Cambridge, MA: Harvard University Press, 1991), 97–100.

14. David Andress, *The Terror* (New York: Little, Brown, 2005), 310–11.

15. Jacques Mallet du Pan, *Considerations sur la nature de la revolution de France* (1793).

16. Mary Pickering, *Auguste Comte*. Vol. 1, *An Intellectual Biography* (Cambridge: Cambridge University Press, 2006), 392.

17. Harriet Martineau, *The Positive Philosophy of Auguste Comte* (London: George Bell and Sons, 1896), Introduction.

18. Martineau, *The Positive Philosophy of Auguste Comte*, Introduction.

19. Martineau, *The Positive Philosophy of Auguste Comte*, Introduction.

20. Martineau, *The Positive Philosophy of Auguste Comte*, Introduction.

21. Pickering, *Auguste Comte*, 545.

22. Phyllis Blanchard, "A Psycho-Analytic Study of Auguste Comte," *American Journal of Psychology* 29, no. 2 (April 1918): 159–81.

23. Tony Davies, *Humanism, The New Critical Idiom* (London and New York: Routledge, 1997), 29

24. J. Peder Zane, *The Top Ten: Writers Pick Their Favorite Books* (New York: W. W. Norton, 2010).

25. Bernard J. Paris, "George Eliot's Religion of Humanity," *ELH* 29, no. 4 (December 1962): 418–43 (emphasis added).

26. Paris, "George Eliot's Religion of Humanity."

27. Michael W. Taylor, *The Philosophy of Herbert Spencer* (New York: Continuum, 2007), 4.

28. Thomas Eriksen and Finn Nielsen, *A History of Anthropology* (London: Pluto, 2013), 37.

29. Robert Bates Graber, "Herbert Spencer and George Eliot: Some Corrections and Implications," *George Eliot—George Henry Lewes Studies* 22/23 (September 1993): 72.

30. Herbert Spencer, *Social Statics* (London: John Chapman, 1851), 65 (emphasis added).

31. Herbert Spencer, *An Autobiography* (New York: D. Appleton & Co., 1904), 2:568.

32. Herbert Spencer, "The Development Hypothesis," *The Leader*, March 20, 1852.

33. T. H. Huxley, "The Origin of Species," in *Darwiniana* (New York: D. Appleton & Co. 1896), 52.

34. Bernard Lightman, "Huxley and Scientific Agnosticism: The Strange History of a Failed Rhetorical Strategy," *British Journal for the History of Science* 35, no. 3 (September 2002): 271–89.

35. Hardin, Numbers, and Binzley, *The Warfare Between Science & Religion*, 65.

36. Robert Bruce Mullin, "Science, Miracles, and the Prayer-Gauge Debate," in *When Science & Christianity Meet*, ed. David C. Lindberg and Ronald L. Numbers (Chicago: University of Chicago Press, 2003), 203–24.

37. John Tyndall, *Address Delivered Before the British Association Assembled at Belfast: With Additions* (London: Longmans, Green and Co., 1874), 2.

38. Tyndall, *Address*, 11.

39. Tyndall, *Address*, 18–19.

40. Tyndall, *Address*, 61 (emphasis added).

41. Tyndall, *Address*, preface.

42. Hardin, Numbers, and Binzley, *The Warfare Between Science & Religion*, 77.

43. Frank M. Turner, *Contesting Cultural Authority: Essays in Victorian Cultural Life* (Cambridge: Cambridge University Press, 1993), 196.

44. *Journals of T. A. Hirst*, November 6, 1864, Tyndall Papers, Royal Institution of Great Britain, London, 5/B4 (emphasis added).

45. Ruth Barton, " 'Huxley, Lubbock, and Half a Dozen Others': Professionals and Gentlemen in the Formation of the X Club, 1851–1864," *Isis* 88 (1998): 410–44. See also her most recent *The X Club: Power and Authority in Victorian Science* (Chicago: University of Chicago Press, 2018).

46. Hardin, Numbers, and Binzley, *The Warfare Between Science & Religion*, 75.

47. Roland Jackson, "John Tyndall in America," *OUPblog*, May 5, 2018, https://blog.oup.com/2018/05/john-tyndall-america/. See his *The Ascent of John Tyndall: Victorian Scientists, Mountaineer, and Public Intellectual* (Oxford: Oxford University Press, 2018).

48. Hardin, Numbers, and Binzley, *The Warfare Between Science & Religion*, 73.

49. Quoted in James C. Ungureanu, "Tyndall and Draper," *Notes and Queries* 64, no. 1 (2017): 127.

50. Hardin, Numbers, and Binzley, *The Warfare Between Science & Religion*, 14 (emphasis added).

51. On White's extensive library, see James C. Ungureanu, *Science, Religion, and the Protestant Tradition: Retracing the Origins of Conflict* (Pittsburgh: University of Pittsburgh Press, 2019), 78.

52. Hardin, Numbers, and Binzley, *The Warfare Between Science & Religion*, 79 (emphasis added).

# Chapter 3

1. *Road Trippin'*, Episode 7, "Kyle Irving," February 16, 2017.

2. *Road Trippin'*.

3. Alfred Russell Wallace, "Reply to Mr Hampden's Charges against Mr Wallace," *The Pamphlet Collection of Sir Robert Stout*, 22:7. http://nzetc.victoria.ac.nz/tm/scholarly/tei-corpus-Stout.html

4. Hoang Nguyen, "Most Flat Earthers Consider Themselves Very Religious," YouGov, April 2, 2018, https://today.yougov.com/topics/philosophy/articles-reports/2018/04/02/most-flat-earthers-consider-themselves-religious.

5. http://www.theflatearthsociety.org/home/index.php/about-the-society/faq.

6. Mark Lamoureux, "Inside Canada's First-Ever Flat Earth Conference," *Vice*, August 30, 2018.

7. Nguyen, "Most Flat Earthers."

8. http://www.theflatearthsociety.org/home/index.php/about-the-society/faq.

9. Christine Garwood, *Flat Earth: The History of an Infamous Idea* (New York: Macmillan, 2010), 114–15.

10. White, *A History of the Warfare*, 1:89.

11. White, *A History of the Warfare*, 1:91.

12. White, *A History of the Warfare*, 1:91.

13. White, *A History of the Warfare*, 1:92.

14. White, *A History of the Warfare*, 1:93.

15. White, *A History of the Warfare*, 1:95, 1:326.

16. White, *A History of the Warfare*, 1:97.

17. White, *A History of the Warfare*, 1:108–109.

18. White, *A History of the Warfare*, 1:109.

19. White, *A History of the Warfare*, 1:109.

20. White, *A History of the Warfare*, 1:109.

21. Draper, *History of the Conflict*, 152.

22. Draper, *History of the Conflict*, 64.

23. Draper, *History of the Conflict*, 64, 65.

24. Draper, *History of the Conflict*, 66.

25. Draper, *History of the Conflict*, 164–65.

26. Daniel J. Boorstin, *The Discoverers* (London: Dent, 1984), xv.

27. Boorstin, *The Discoverers*, 107.

28. Boorstin, *The Discoverers*, 107.

29. Boorstin, *The Discoverers*, 108.

30. John D. Fix, *Astronomy: Journey to the Cosmic Frontier*, 6th ed, (New York: McGraw-Hill, 2011), 58 (thanks to Michael Newton Keas for finding this).

31. Edward Grant, *God and Reason in the Middle Ages* (Cambridge: Cambridge University Press, 2001), 343.

32. https://twitter.com/neiltyson/status/692939759593865216.

33. https://twitter.com/neiltyson/status/692939759593865216.

34. Draper, *History of the Conflict*, 294.

35. Jeffrey Burton Russell, *Inventing the Flat Earth* (Westport, CT: Praeger, 1991), xii.

36. Russell, *Inventing the Flat Earth*, xii.

37. *The Myth of the Flat Earth*. American Scientific Affiliation. https://www.asa3.org/ASA/topics/history/1997Russell.html

38. Allan Chapman, *Slaying the Dragons* (Oxford: Lion, 2013), 60–61.

39. Lesley B. Cormack, *Medieval Christianity and the Flat Earth*, in *Galileo Goes to Jail*, ed. Ronald L. Numbers (Cambridge, MA: Harvard University Press, 2009), 30–31.

40. Cormack, *Medieval Christianity and the Flat Earth*, 61.
41. White, *A History of the Warfare*, 1:113.
42. Lactantius, *Ante-Nicene Fathers*, 7:157–58.
43. Patrick Healy, "Lucius Caecilius Firmianus Lactantius," in *The Catholic Encyclopedia*, vol. 8 (New York: Robert Appleton Company, 1910).
44. *Encyclopedia Britannica Micropedia*, 15th ed., 1993, 7:90. The original source is Letter LVIII to Paulinus, paragraph 10, by St. Jerome.
45. Pablo de Felipe and Malcolm A. Jeeves, *Science and Christianity Conflicts: Real and Contrived*, Perspectives on Science and Christian Faith 69, no. 3 (September 2017): 131–47.
46. Russell, *Inventing the Flat Earth*, 34–35.
47. Philoponus, *De Opificio Mundi*, 3:8, trans. L. S. B. MacCoull, 106, 117.
48. Pablo de Felipe and Robert D. Keay, "Science and Faith Issues in Ancient and Medieval Christianity," BioLogos, December 2, 2013, https://biologos.org/articles/science-and-faith-issues-in-ancient-and-medieval-christianity.
49. Darin Hayton, "Washington Irving's Columbus and the Flat Earth," December 2, 2014, https://dhayton.haverford.edu/blog/2014/12/02/washington-irvings-columbus-and-the-flat-earth/.
50. Washington Irving, *A History of the Life and Voyages of Christopher Columbus*, 4 Vols. (London: John Murray, 1828), 1:121.
51. Draper, *History of the Conflict*, 104.
52. Hayton, "Washington Irving's Columbus."
53. Russell, *Inventing the Flat Earth*, 46.
54. Maya Shwayder, "Debunking a Myth," *Harvard Gazette*, April 7, 2011.

# Chapter 4

1. Patrick M. Grant, et al., "A Possible Chemical Explanation for the Events Associated with the Death of Gloria Ramirez at Riverside General Hospital," *Forensic Science International* 87, no. 3 (1997): 219–37.
2. Edmund S. Meltzer and Gonzalo M. Sanchez, *The Edwin Smith Papyrus: Updated Translation of the Trauma Treatise and Modern Medical Commentaries*, ISD LLC, June 23, 2014, 53.
3. Heinrich Von Staden, "The Discovery of the Body: Human Dissection and Its Cultural Contexts in Ancient Greece," *Yale Journal of Biology and Medicine* 65 (1992): 223–41.
4. White, *A History of the Warfare*, 2:1.
5. White, *A History of the Warfare*, 2:2.
6. White, *A History of the Warfare*, 2:3.
7. See Matthew 25:31–45.
8. White, *A History of the Warfare*, 2:90.
9. White, *A History of the Warfare*, 2:31.
10. White, *A History of the Warfare*, 2:31–32.

11. White, *A History of the Warfare*, 2:50.

12. White, *A History of the Warfare*, 2:54.

13. Allan Chapman, *Physicians, Plagues and Progress* (Oxford: Lion, 2016), 395.

14. "Proposed Memorial to Robert Liston," *British Medical Journal* 2, no. 2483 (August 1, 1908): 284–85.

15. J. Y. Simpson, "Answer to the Religious Objections Advanced Against the Employment of Anaesthetic Agents in Midwifery and Surgery," reprinted in *British Journal of Anaesthesia* 31, no. 1 (January 1959): 35–43.

16. Simpson, "Answer to the Religious Objections."

17. Draper, *History of the Conflict*, 318–19.

18. Draper, *History of the Conflict*, 318.

19. White, *A History of the Warfare*, 2:55.

20. Draper, *History of the Conflict*, 269.

21. Draper, *History of the Conflict*, 269.

22. White, *A History of the Warfare*, 2:37.

23. Draper, *History of the Conflict*, 270.

24. Draper, *History of the Conflict*, 270.

25. White, *A History of the Warfare*, 2:38–39.

26. James Hannam, *God's Philosophers* (London: Icon, 2009), 1693, 106.

27. Walter Clyde Curry, *Chaucer and the Medieval Sciences* (New York: Barnes and Noble, 1960), xi.

28. J. H. G. Grattan and Charles Singer, *Anglo-Saxon Magic and Medicine* (London: Oxford University Press, 1952), 92.

29. "Vesalius's Renaissance Anatomy Lessons," British Library, http://vll-minos.bl.uk/learning/cult/bodies/vesalius/renaissance.html.

30. Jeffrey A. Norton et al., eds., *Surgery: Basic Science and Clinical Evidence*, 2nd ed. (New York: Springer Science & Business Media, 2009), 6.

31. The website, to the great credit of the BBC, has been substantially updated following several complaints about its inaccurate content. Its current wording is far more in keeping with the latest scholarship, and can be found at https://www.bbc.co.uk/bitesize/guides/zxg6wxs/revision/1.

32. Speech in favor of the Alternative Pluripotent Stem Cell Research Enhancement Act, July 21, 2005.

33. C. Howse, "The Myth of the Anatomy Lesson," *Telegraph*, June 10, 2009.

34. Heinrich Von Staden, "The Discovery of the Body: Human Dissection and Its Cultural Contexts in Ancient Greece," *Yale Journal of Biology and Medicine* 65 (1992): 223–41.

35. Katharine Park, "That the Medieval Church Prohibited Human Dissection," in *Galileo Goes to Jail*, ed. Ronald L. Numbers (Cambridge, MA: Harvard University Press, 2009), 44.

36. C. H. Talbot, *Medicine in Medieval England* (London: Oldbourne, 1967), 55.

37. Talbot, *Medicine in Medieval England*.

38. Darrel W. Amundsen, "Medieval Canon Law on Medical and Surgical Practice by the Clergy," *Bulletin of the History of Medicine* 52, no. 1 (1978): 22–44.

39. Andrew Cunningham, *The Anatomist Anatomis'd: An Experimental Discipline in Enlightenment Europe* (Farnham, UK: Ashgate, 2010), 14.
40. Park, *That the Medieval Church*, 47.
41. Park, *That the Medieval Church*, 47.
42. Cunningham, *The Anatomist Anatomis'd*, 13.
43. As quoted by James Hannam in *The Deep Sleep of Adam*, *Quodlibeta*, December 19, 2008. url: https://bedejournal.blogspot.com/2008/12/deep-sleep-of-adam.html
44. See discussion in Chapman, *Physicians, Plagues and Progress*, 404–05.
45. Maxine Van De Wetering, "A Reconsideration of the Inoculation Controversy," *New England Quarterly* 58, no. 1 (March 1985): 46–67.
46. M. Best, D. Neuhauser, and L. Slavin, " 'Cotton Mather, You Dog, Dam You! I'l Inoculate You with This; with a Pox to You': Smallpox Inoculation, Boston, 1721," *Quality and Safety in Health Care* 13 (2004): 82–83.
47. Sanjib Kumar Ghosh, "Human Cadaveric Dissection: A Historical Account from Ancient Greece to the Modern Era," *Anatomy & Cell Biology* 48, no. 3 (2015): 153–69.
48. Shwayder, "Debunking a Myth."
49. M. L. Cameron, *Anglo-Saxon Medicine* (Cambridge: Cambridge University Press, 1993), 3.
50. Richard Raiswell, "The Age Before Reason," in *Misconceptions About the Middle Ages*, ed. Stephen J. Harris and Bryon L. Grigsby (New York and London: Routledge, 2008), 125.

# Chapter 5

1. https://www.theodysseyonline.com/top-ten-list-of-top-ten-top-ten-lists.
2. Alex Bellos, "The 10 Best Mathematicians," *The Guardian*, April 11, 2010.
3. Bellos, "The 10 Best Mathematicians."
4. Edward Gibbon, "Ecclesiastical Discord Pt. II," in *Decline and Fall of the Roman Empire*, vol. 4, chap. XLVII (1788).
5. Catherine Nixey, *The Darkening Age: The Christian Destruction of the Classical World* (Boston and New York: Houghton Mifflin Harcourt, 2018), 88.
6. Nixey, *The Darkening Age*, 136.
7. Draper, *History of the Conflict*, 54–56.
8. David C. Lindberg, "That the Rise of Christianity Was Responsible for the Demise of Ancient Science," in *Galileo Goes to Jail*, ed. Ronald L. Numbers (Cambridge, MA: Harvard University Press, 2009), 9.
9. As given in Charles Freeman, *The Closing of the Western Mind* (London: Pimlico, 2003), in the front matter.
10. As given in Freeman, *The Closing of the Western Mind*, front matter.
11. Draper, *History of the Conflict*, 62.
12. White, *A History of the Warfare*, 1:374, 375.
13. White, *A History of the Warfare*, 1:375.

14. Draper, *History of the Conflict*, 157–158.
15. White, *A History of the Warfare*, 1:376.
16. White, *A History of the Warfare*, 1:32.
17. White, *A History of the Warfare*, 1:33.
18. Dennis D. McManus, *Isidore of Seville: De Ecclesiasticis Officiis* (Mahwah, NJ: Paulist Press, 2008), 11.
19. Cosmas Indicopleustes, *The Christian Topography of Cosmas, An Egyptian Monk* (London: Hakluyt Society, 1897), 360–361.
20. White, *A History of the Warfare*, 1:38.
21. Draper, *History of the Conflict*, 268–269.
22. Winston Black, *The Middle Ages: Facts and Fictions* (Santa Barbara, CA: ABC-CLIO, 2019), 112.
23. White, *A History of the Warfare*, 1:25.
24. White, *A History of the Warfare*, 2:23.
25. White, *A History of the Warfare*, 2:23.
26. Janet L. Nelson, "The Dark Ages," *History Workshop Journal* 63 (Spring 2007): 191–201, on 193 (my emphasis).
27. Nelson, "The Dark Ages."
28. Program info and clips available at https://www.bbc.co.uk/programmes/b00z8r9l.
29. Full details of the program at https://www.pbs.org/wgbh/nova/archimedes/about.html.
30. Draper, *History of the Conflict*, 215, 264–65, 307, 255.
31. White, *A History of the Warfare*, 1:376.
32. Carl Sagan, *Cosmos* (New York: Random House, 1980). 335.
33. Michael H. Shank, "That the Medieval Christian Church Suppressed the Growth of Science," in *Galileo Goes to Jail* (Cambridge, MA: Harvard University Press, 2009). 20.
34. Seb Falk, *The Light Ages* (New York: W. W. Norton, 2020), 78.
35. Falk, *The Light Ages*, 6.
36. O. M. Bakke, *When Children Became People*, trans. Brian McNeil (Minneapolis: Augsburg Fortress, 2005), 3208–09.
37. Tertullian, "Of Schoolmasters and Their Difficulties," ANF, vol. 3, part 1, chap. 10.
38. Hypatia was not an atheist, but a neo-Platonist. She was one of many women in prominent positions. Her relationships with Christians were generally good. See her biography—Maria Dzielska, *Hypatia of Alexandria* (Cambridge, MA: Harvard University Press, 1996).
39. Personal correspondence, June 6, 2020. Readers may be interested in Dickson's book, Bullies and Saints: An Honest Look at the Good and Evil of Church History, Zondervan Academic, 2021—which contains a chapter on Hypatia
40. Michael A. B. Deakin, "Hypatia and Her Mathematics," *American Mathematical Monthly* 101, no. 3 (March 1994), 234–43.
41. David Bentley Hart, "The Perniciously Persistent Myths of Hypatia and the Great Library," *First Things*, June 4, 2010, https://www.firstthings.com/web-exclusives/2010/06/the-perniciously-persistent-myths-of-hypatia-and-the-great-library.
42. Hart, "The Perniciously Persistent Myths."

43. See, for instance, Tim O'Neill's collection of comments about the library, "The Great Myths 5: The Destruction of the Great Library of Alexandria," History for Atheists, July 2, 2017, https://historyforatheists.com/2017/07/the-destruction-of-the-great-library-of-alexandria/.

44. Sagan, *Cosmos*, 363.

45. Roger S. Bagnall, "Alexandria: Library of Dreams," *Proceedings of the American Philosophical Society* 146, no. 4 (December 2002): 348–62.

46. Bagnall, "Alexandria: Library of Dreams."

47. Bagnall, "Alexandria: Library of Dreams."

48. A. C. Grayling, *The History of Philosophy* (New York: Penguin, 2020), 3.

49. Tom Holland and A. C. Grayling, "History: Did Christianity Give Us Our Human Values?," *The Big Conversation Show*, episode 5, season 2, https://www.premierchristianradio.com/Shows/Saturday/Unbelievable/Episodes/Unbelievable-Tom-Holland-vs-AC-Grayling-Did-Christianity-give-us-our-human-values

50. Holland and Grayling, "History."

51. Holland and Grayling, "History."

52. James Hannam, *God's Philosophers* (London: Icon, 2009), 80, 117.

53. Christian Wildberg, "John Philoponus," *The Stanford Encyclopedia of Philosophy* (Winter 2018 Edition), ed. Edward N. Zalta, https://plato.stanford.edu/archives/win2018/entries/philoponus/.

54. Allan Chapman, *Gods in the Sky* (London: Channel 4 Books, 2002), 173

55. Bede, *On the Nature of Things and on Times*, trans. Calvin B. Kendall and Faith Wallis (Liverpool: Liverpool University Press, 2010).

56. John Hadley and David Singmaster, "Problems to Sharpen the Young," *Mathematical Gazette* 76, no. 475 (March 1992), 102–26.

57. George Ovitt, Jr., "Technology and Science," in *The Cambridge History of Science*, vol. 2—*Medieval Science* (Cambridge: Cambridge University Press, 2013), 630–44.

58. Ovitt, "Technology and Science."

59. Peter Dendle, "The Middle Ages Were a Superstitious Time," in *Misconceptions About the Middle Ages*, ed. Stephen J. Harris and Bryon L. Grigsby (New York and London: Routledge, 2008), 117–18.

60. Dendle, "The Middle Ages."

61. BBC News, "Gwyneth Paltrow's Goop Pays $145,000 in Vaginal Egg Lawsuit," September 5, 2018, https://www.bbc.co.uk/news/world-us-canada-45426332.

62. "The Energy Experience," *the goop lab*, episode 5, aired January 24, 2020.

63. *Physics Professor watches "the goop lab"*—*Sixty Symbols*, YouTube, Februay 11, 2020, https://www.youtube.com/watch?v=EIyQcGyRXwg.

64. Kate Clark, "Gwyneth Paltrow's Goop Raises $50M at $250M Valuation," PitchBook, March 28, 2018, https://pitchbook.com/news/articles/gwyneth-paltrows-goop-raises-50m-at-250m-valuation.

65. Dendle, "The Middle Ages."

66. Annemarie de Waal Malefijt, "Homo Monstrosus," *Scientific American* 219, no. 4 (October 1968): 112–19.

67. Isaac Asimov, *Asimov's Guide to the Bible the Old and New Testaments*, 2 vols. (Avenel, NJ: Wings Books, 1981), 186–87.

68. See J. Gerard, *Of a Bull and a Comet* (Philadelphia: American Ecclesiastical Review, 1908) and William P. Rigge, "The Pope and the Comet," *Popular Astronomy* 16 (1908): 481–83.

69. Hannam, *God's Philosophers*, 1854.

70. Hannam, *God's Philosophers*, 1502.

71. Black, *The Middle Ages*, 179.

72. Bernard of Clairvaux, *Apologia*, trans. David Burr. https://sourcebooks.fordham.edu/source/bernard1.asp

73. Bakke, *When Children Became People*, 355.

74. Charles Singer et al., *A History of Technology*, vol 2 (Oxford: Clarendon Press, 1956), 690.

75. Dolly Jørgensen, "Modernity and Medieval Muck," *Nature and Culture* 9, no. 3 (Winter 2014): 225–37.

76. *The City of God*, trans. Philip Schaff (New York: Christian Literature Publishing, 1890), 129–40.

77. *The City of God*, 529–30.

78. *The City of God*, 648.

79. Mikhail, "Alternavox at TIFF: In Conversation with Alejandro Amenábar," Deadline, September 18, 2009, http://www.alternavox.net/6875/alternavox-at-tiff-in-conversation-withalejandro-amenabar/.

80. Ovitt, "Technology and Science."

81. Ovitt, "Technology and Science."

# Chapter 6

1. BBC News, "Yasaku Maezawa: Japanese Billionaire Seeks 'Life Partner' for Moon Voyage," January 13, 2020, https://www.bbc.co.uk/news/world-asia-51086635.

2. BBC News, "Yasaku Maezawa."

3. BBC News, "Yasaku Maezawa." The reader may be interested to know that Maezawa retracted the offer in late January 2020, and has replaced it with a new one: he will now take eight people with him instead.

4. White, *A History of the Warfare*, 1:57.

5. "Standing Up in the Milky Way," episode 1 of *Cosmos: A Spacetime Odyssey*, 20th Century Fox Home Entertainment, aired March 9, 2014.

6. "Standing Up in the Milky Way."

7. "Standing Up in the Milky Way."

8. "Standing Up in the Milky Way."

9. Draper, *History of the Conflict*, 181.

10. Giordano Bruno, *On the Infinite, the Universe, & the Worlds*, trans. Scott Gosnell (London: Huginn, Munnin & Co.), 29.

11. See James Hannam, *God's Philosophers* (London: Icon, 2009), 1550.
12. Bruno, *On the Infinite, the Universe, & the Worlds*, 110.
13. For a blow-by-blow account of Bruno's mistakes, see Giordano Bruno, *The Ash Wednesday Supper (La Cena de le Ceneri)*, trans. with intro. and notes, Stanley L. Jaki (Paris: Mouton, 1584), https://math.dartmouth.edu/~matc/Readers/renaissance.astro/6.1.Supper.html.
14. Maurice A. Finocchiaro, "Review of Giordano Bruno: An Introduction by Blum and Henneveld," *Isis* 105, no. 3 (September 2014): 631–32.
15. Bruno, *On the Infinite, the Universe, & the Worlds*, 139.
16. Finocchiaro, "Review of Giordano Bruno."
17. See, for instance, Edward Grant, "Jean Buridan and Nicole Oresme on Natural Knowledge," *Vivarium* 31, no. 1 (1993): 84–105.
18. Jole Shackleford, "That Giordano Bruno Was the First Martyr of Modern Science," in *Galileo Goes to Jail*, ed. Ronald L. Numbers (Cambridge, MA: Harvard University Press, 2009), 65.
19. "Standing Up in the Milky Way."
20. Open Yale Courses, "HIST 202: European Civilization, 1648–1945, Lecture 4—Peter the Great," video, 12 minute mark, https://oyc.yale.edu/history/hist-202/lecture-4.
21. White, *A History of the Warfare*, 1:132.
22. Draper, *History of the Conflict*, 171–72.
23. Galileo Galilei, "Letter to the Grand Duchess Christina of Tuscany, 1615," English text available at https://sourcebooks.fordham.edu/mod/galileo-tuscany.asp.
24. Galilei, "Letter to the Grand Duchess."
25. Thomas M. Lessl, "The Galileo Legend as Scientific Folklore," *Quarterly Journal of Speech* 85, no. 2 (1999): 146–68.
26. Allan Chapman, *Slaying the Dragons* (Oxford: Lion, 2013), 107.
27. Lessl, "The Galileo Legend."
28. Mike Brown, "'A More Perfect Heaven,' by Dava Sobel, Is About the Revolutionary Idea of Copernicus," *Washington Post*, October 14, 2011.
29. Brown, "'A More Perfect Heaven.'"
30. Owen Gingerich, *God's Planet* (Cambridge, MA: Harvard University Press, 2014), 12,13.
31. White, *A History of the Warfare*, 1:121–22.
32. See discussion in Gingerich, *God's Planet*.
33. Bruno, "Of the Second Proposition of Nundinio," in *The Ash Wednesday Supper*.
34. Robert S. Westman, "The Copernicans and the Churches," in *God and Nature: Historical Essays on the Encounter Between Christianity and Science*, eds. David C. Lindberg and Ronald L. Numbers (Berkeley: University of California Press, 1986), 103.
35. White, *A History of the Warfare*, 1:127.
36. The closest Calvin comes is when he denounces those who deliberately run contrary to reason "like the man who said snow is black" or "it is the earth which shifts and turns." Robert White, in "Calvin and Copernicus: The Problem Reconsidered," *Calvin Theological Journal* 15 (1980), 233–43, at 236–37.

37. Edward Rosen, "Calvin's Attitude Toward Copernicus," *Journal of the History of Ideas* 21, no. 3 (July–September 1960): 431–41.

38. Rosen, "Calvin's Attitude Toward Copernicus."

39. Calvin, "Sermons on the book of Job," in *Calvini Opera*, vol. 34, 429–30.

40. Morris Kline, *Mathematics and the Search for Knowledge* (New York: Oxford University Press, 1985), 72.

41. Michael Newton Keas, *Unbelievable: 7 Myths About the History and Future of Science and Religion* (Wilmington, DE: ISI Books, 2018).

42. Dennis R. Danielson, "That Copernicanism Demoted Humans from the Center of the Cosmos," in *Galileo Goes to Jail*, 55.

43. Lessl, "The Galileo Legend."

44. https://whyevolutionistrue.wordpress.com/2017/01/21/yet-another-accommodationist-book/.

45. Jerry A. Coyne, *Faith vs. Fact: Why Science and Religion Are Incompatible* (New York: Penguin, 2015), xii.

46. Thomas Williams, "Saint Anselm," in *Stanford Encyclopedia of Philosophy* (Spring 2016 Edition), ed. Edward N. Zalta, https://plato.stanford.edu/archives/spr2016/entries/anselm/.

47. Anselm, *Complete Philosophical and Theological Treatises of Anselm of Canterbury*, ed. and trans. Jasper Hopkins and Herbert Richardson (Minneapolis: Arthur J. Banning Press, 2000).

48. See, for instance, Günther Eder and Esther Ramharter, "Formal Reconstructions of St. Anselm's Ontological Argument," *Synthese* 192, no. 9 (2015): 2795–825.

49. Richard Raiswell, "The Age Before Reason," in *Misconceptions About the Middle Ages*, ed. Stephen J. Harris and Bryon L. Grigsby (New York and London: Routledge, 2008), 129.

50. Latin text available at http://individual.utoronto.ca/pking/resources/abelard/Sic_et_non.txt.

51. Full text available at https://archive.org/details/DivineBenevolenceOrAnAttemptTo ProveThatThe/mode/2up.

52. See, for instance, Andrzej Grzegorczyk, "Undecidability without Arithmetization," *Studia Logica: An International Journal for Symbolic Logic* 79, no. 2 (2005): 163–230.

53. Petr Hájek, "Ontological Proofs of Existence and Non-Existence," *Studia Logica: An International Journal for Symbolic Logic* 90, no. 2 (November 2008): 257–62.

54. https://philpapers.org/surveys/.

55. Ard Louis, "How Do I Obtain Reliable Knowledge About the World?," in *A Teacher's Guide to Science and Religion in the Classroom*, ed. Berry Billingsley, Manzoorul Abedin, and Keith Chappell (London and New York: Routledge, 2018), 150.

56. White, *A History of the Warfare*, 1:387.

57. Ronald L. Numbers, "Introduction," in *Galileo Goes to Jail*, 6.

58. Michael Brooks, *The Secret Anarchy of Science* (London: Profile, 2012), 2.

59. Tom McLeish, *Faith and Wisdom in Science* (Oxford: Oxford University Press, 2014), 199–200.

60. See, for example, *Science Fictions* by Stuart Ritchie; *Bad Science* by Ben Goldacre; *Lost in Math* by Sabine Hossenfelder; *Faster Than the Speed of Light* by Joao Magueijo; *The Trouble with Physics* by Lee Smolin.

61. Henry M. Cowles, *The Scientific Method* (Cambridge, MA: Harvard University Press, 2020), 1, 2.

62. Draper, *History of the Conflict*, 33.

63. Sean Carroll, *Something Deeply Hidden* (London: Oneworld, 2019), 178.

64. Everett to Petersen, May 31, 1957, Wheeler Papers, Series I—Box Di—Fermi Award #1—Folder Everett, quoted in Stefano Osnaghi, Fabio Freitas, and Olival Freire Jr., "The Origin of the Everettian Heresy," *Studies in History and Philosophy of Science Part B Studies in History and Philosophy of Modern Physics* 40 (2009): 97–123.

65. J. S. Bell, *Speakable and Unspeakable in Quantum Mechanics* (Cambridge: Cambridge University Press, 2004), 160.

66. Adam Becker, *What Is Real?: The Unfinished Quest for the Meaning of Quantum Physics* (London: John Murray, 2018), 3950.

67. When Scientific Orthodoxy Resembles Religious Dogma, Avi Loeb, Scientific American, May 17, 2021 url: https://www.scientificamerican.com/article/when-scientific-orthodoxy-resembles-religious-dogma/

68. Becker, *What Is Real?*, 4550.

69. BBC Sport, "Azerbaijan Grand Prix: The Secret Aerodynamicist on Design Icon Adrian Newey," April 26, 2019, https://www.bbc.co.uk/sport/formula1/47838557.

70. See Brooks, *The Secret Anarchy of Science*, chapter 1, for the full accounts.

71. Coyne, *Faith vs. Fact*, xii.

# Chapter 7

1. Frank M. Turner, "The Victorian Conflict Between Science and Religion: A Professional Dimension," *Isis* 69, no. 3 (1978): 356–76.

2. Turner, *The Victorian Conflict*.

3. Turner, *The Victorian Conflict*.

4. Draper, *History of the Conflict*, 284.

5. Donald Fleming, *John William Draper and the Religion of Science* (London: Octagon, 1972), 31.

6. White, *A History of the Warfare*, 2:vii–viii.

7. Fleming, *John William Draper*, 31.

8. White, *A History of the Warfare*, 2:394.

9. Draper, *History of the Conflict*, 364.

10. Francis Bacon, "The Advancement of Learning," in *The Major Works*, ed. Brian Vickers (New York: Oxford University Press, 2008), 138, 141–42, 152–53.

11. See James Hannam, *God's Philosophers* (London: Icon, 2009), for plenty of examples of mathematical science pre-Bacon. The purge mentioned happened under the watch

of Thomas Cromwell (1485–1540)—as Hannam wrote in his PhD thesis, "Teaching Natural Philosophy and Mathematics at Oxford and Cambridge 1500–1570."

12. See Bacon, "New Atlantis," in *The Major Works*, ed. Vickers, 457–89.

13. Peter Gay, *The Enlightenment: An Interpretation* (New York and London: W. W. Norton, 1995), 1:327.

14. Quoted in James C. Ungureanu, *Science, Religion, and the Protestant Tradition: Retracing the Origins of Conflict* (Pittsburgh: University of Pittsburgh Press, 2019), 130.

15. Ungureanu, *Science, Religion, and the Protestant Tradition*.

16. Ungureanu, *Science, Religion, and the Protestant Tradition*.

17. Draper, *Introductory Lecture, to the Course of Chemistry: On the Relations and Nature of Water* (New York: New York University, 1845–1846), 5, 9, 13.

18. Draper, *History of the Conflict*, 225.

19. Draper, *Introductory Lecture on Oxygen Gas* (New York: Joseph H. Jennings, 1848), 6.

20. *Catalogue of the Historical Library of Andrew Dickson White*, vol. 1: *The Protestant Reformation* (Ithaca: Cornell University Press, 1889).

21. Robert Morris Ogden, ed., *The Diaries of Andrew D. White* (Ithaca: Cornell University Press, 1959), 79.

22. Quoted in Ungureanu, *Science, Religion, and the Protestant Tradition*, 79.

23. Karl Barth, *Protestant Theology in the Nineteenth Century* (Grand Rapids: Eerdmans, 1959), 440.

24. Quoted in Ungureanu, *Science, Religion, and the Protestant Tradition*, 80.

25. White, *A History of the Warfare*, 2:393–94.

26. White, *A History of the Warfare*, 2:395–96. (Emphasis added).

27. For a much fuller analysis of the reception of their work, see Ungureanu, *Science, Religion*, 216–48.

28. *New York Evangelist*, quoted in Ungureanu, *Science, Religion*, 238.

29. Quoted in Ungureanu, *Science, Religion*, 224–25.

30. Quoted in Ungureanu, *Science, Religion*, 238.

31. Quoted in Ungureanu, *Science, Religion*, 238.

32. Quoted in Ungureanu, *Science, Religion*, 238.

33. Quoted in Ungureanu, *Science, Religion*, 239.

34. Quoted in Ungureanu, *Science, Religion*, 239.

35. Quoted in Ungureanu, *Science, Religion*, 240.

# Chapter 8

1. Draper, *History of the Conflict*, 218.

2. White, *A History of the Warfare*, 1:113.

3. Harry Kroto, "Blinded by a Divine Light," *The Guardian*, September 28, 2008.

4. Friedrich Nietzsche, *On the Genealogy of Morals and Ecce Homo*, trans. Walter Kaufmann (New York: Vintage, 1967), 151–52.

5. Peter Harrison, "Religion and the Early Royal Society," *Science and Christian Belief* 22 (2010): 3–22.

6. Of course, other faiths and philosophies are also monotheistic; discussion of this, however, is beyond the scope of this particular volume.

7. Mark Worthing, "Monotheism and the Origins of Natural Science," http://www. iscast.org/resources/Worthing_M_2017-12-Monotheism_and_the_Origins_of_ Natural_Science.pdf.

8. See, for instance, J. C. O'Neill, "How Early Is the Doctrine of *Creatio ex Nihilo*?," *Journal of Theological Studies*, New Series 53, no. 2 (October 2002): 449–65.

9. Johannes Kepler, *Optics*, trans. William H. Donahue (Santa Fe, NM: Green Lion Press, 2000), 15.

10. Yahuda MS 1.1, National Library of Israel, *Untitled Treatise on Revelation* (section 1.1): http://www.newtonproject.ox.ac.uk/view/texts/normalized/THEM00135.

11. Robert Boyle, *A Free Enquiry into the Vulgarly Received Notion of Nature*, ed. Edward B. Davis and Michael Hunter (Cambridge: Cambridge University Press, 1996), 24.

12. Ian H. Hutchinson, "The Genius and Faith of Faraday and Maxwell," *The New Atlantis* 41 (Winter 2014): 81–99.

13. John C. Lennox, *God's Undertaker* (Oxford: Lion, 2009), 63.

14. Picard, "Newton—Rationalizing Christianity, or Not?" (MIT lecture, 1997), full transcript available at https://web.media.mit.edu/~picard/personal/Newton.php.

15. John Polkinghorne, *Science & Theology* (Minneapolis: Fortress, 1998), 73.

16. Rebecca McLaughlin, *Confronting Christianity* (Wheaton, IL: Crossway, 2019), 130.

17. Draper, *History of the Conflict*, 107.

18. White, *A History of the Warfare*, 1:15.

19. Edward Grant, "Reflections of a Troglodyte Historian of Science," *Osiris* 27, no. 1 (2012): 133–55.

20. Grant, "Reflections of a Troglodyte."

21. Grant, "Reflections of a Troglodyte."

22. Grant, "Reflections of a Troglodyte."

23. R. J. Berry, ed., *Real Scientists, Real Faith* (Mansfield, TX: Monarch, 2009), 267.

24. Descartes, *Discourse on the Method*, trans. F. E. Sutcliffe (London: Penguin, 1998), 53. Original work published 1637.

25. *Descartes to Mersenne*, 15 April 15, 1630, *The Philosophical Writings of Descartes*, vol. 3: *The Correspondence*, ed. Cottingham et al. (Cambridge: Cambridge University Press, 1991), 22.

26. Margaret J. Osler, "Whose Ends? Teleology in Early Modern Natural Philosophy," *Osiris* 16 (2001): 151–68.

27. Ted Davis, "The Faith of a Great Scientist: Robert Boyle's Religious Life, Attitudes, and Vocation," BioLogos, August 8, 2013, https://biologos.org/articles/ the-faith-of-a-great-scientist-robert-boyles-religious-life-attitudes-and-vocation.

28. Francis Bacon, "Novum Organum II, §52," in *The Works of Francis Bacon*, 14 vols, James Spedding et al. (London: Longman, 1857–1874), 4:247–48.

29. Harrison, "Religion and the Early Royal Society."

30. Noah Efron, "Christianity Gave Birth to Modern Science," in *Galileo Goes to Jail*, ed. Ronald L. Numbers (Cambridge, MA: Harvard University Press, 2009), 81.

31. See Christopher Insole, "Kant's Transcendental Idealism and Newton's Divine Sensorium," *Journal of the History of Ideas* 72, no. 3 (July 2011): 413–36, for more discussion of Newton and omnipresence.

32. John Polkinghorne, *Science and the Trinity: The Christian Encounter with Reality* (New Haven, CT: Yale University Press, 2008), 61.

33. Peter Barker and Bernard R. Goldstein, "Theological Foundations of Kepler's Astronomy," *Osiris* 16 (2001): 88–113.

34. See Noah Efron's summary in "Christianity Gave Birth to Modern Science."

# Chapter 9

1. Krystie Lee Yandoli, "Here's What It's Like to Visit an Actual Paper Town," Buzzfeed, October 27, 2015, https://www.buzzfeed.com/krystieyandoli/welcome-to-the-agloe-general-store-come-back-soon.

2. Charles M. Haar, "E. L. Youmans: A Chapter in the Diffusion of Science in America," *Journal of the History of Ideas* 9, no. 2 (April 1948): 193–213.

3. For a more detailed study on Youmans and his project, see James C. Ungureanu, "Edward L. Youmans and the 'Peacemakers' in the *Popular Science Monthly*," *Fides et Historia* 51, no. 2 (2019): 13–32.

4. Haar, "E. L. Youmans."

5. George Sarton, "L'Historie de la science," *Isis* 1, no. 1 (1913): 3–46. An English translation of this article was published as "The History of Science," *Monist* 26, no. 3 (1916): 321–65.

6. James C. Ungureanu, *Science, Religion, and the Protestant Tradition: Retracing the Origins of Conflict* (Pittsburgh: University of Pittsburgh Press, 2019), 252.

7. Sarton, "The History of Science," 339.

8. Sarton, "The History of Science," 339.

9. George Sarton, *A Guide to the History of Science: A First Guide for the Study of the History of Science, with Introductory Essays on Science and Tradition* (Waltham, MA: Chronica Botania, 1952), 118.

10. See Ungureanu, *Science, Religion, and the Protestant Tradition*, 249–59.

11. Dan Brown, *Origin* (New York: Penguin, 2017), 109.

12. Brown, *Origin*, 110.

13. Brown, *Origin*, 470.

14. Ronald L. Numbers and Jeff Hardin, "New Atheists," in *The Warfare Between Science and Religion*, ed. Jeff Hardin, Ronald L. Numbers, and Ronald A. Binzley (Baltimore: Johns Hopkins University Press), 220–33.

15. As quoted in Numbers and Hardin, "New Atheists."

16. Numbers and Hardin, "New Atheists."

17. Peter Harrison, "Christianity and the Rise of Western Science," ABC Religion & Ethics, May 8, 2012, https://www.abc.net.au/religion/christianity-and-the-rise-of-western-science/10100570.

18. Alister E. McGrath, *Dawkins' God: Genes, Memes, and the Meaning of Life* (Hoboken, NJ: Wiley, 2013), 142.

19. Ming-Jin Liu et al., "Biomechanical Characteristics of Hand Coordination in Grasping Activities of Daily Living," PLos ONE 11, no. 1 (2016): e0146193, https://doi.org/10.1371/journal.pone.0146193.

20. Liu et al., "Biomechanical Characteristics of Hand Coordination" (under "Reader Comments" section).

21. Daniel Cressey, "Paper That Says Human Hand Was 'Designed by Creator' Sparks Concern," *Nature* 531, no. 7593 (March 3, 2016). https://www.nature.com/news/paper-that-says-human-hand-was-designed-by-creator-sparks-concern-1.19499.

22. Leonid Schneider, "Hand of God Paper Retracted: PLOS ONE 'Could Not Stand by the Pre-Publication Assessment,' " For Better Science, March 4, 2016, https://forbetterscience.com/2016/03/04/hand-of-god-paper-retracted-plos-one-could-not-stand-by-the-pre-publication-assessment/.

23. Liu et al., "Biomechanical Characteristics of Hand Coordination."

24. MingJin Liu, Twitter, July 5, 2016, https://twitter.com/mingjinliu?lang=en.

25. Liu et al., "Biomechanical Characteristics of Hand Coordination."

26. Cressey, "Paper That Says Human Hand."

27. P. Z. Myers, "The Human Hand Is Good at Grasping. Therefore, God." *Pharyngula* (blog), March 2, 2016, https://freethoughtblogs.com/pharyngula/2016/03/02/the-human-hand-is-good-at-grasping-therefore-god/#ixzz41pyh2rYR.

28. Peter J. Boyer, "The Covenant: Francis Collins, a Fervent Christian, Thought He Had Resolved the Stem-Cell Debate. A Federal Judge Disagreed," *New Yorker*, September 6, 2010, https://www.newyorker.com/magazine/2010/09/06/the-covenant.

29. Elaine Howard Ecklund et al., *Secularity and Science: What Scientists Around the World Really Think About Religion* (New York: Oxford University Press, 2019) 46–47.

30. Ecklund et al., *Secularity and Science*, 47.

31. Jerry Coyne, "Evolution and Atheism: Best Friends Forever," Freedom from Religion Foundation, October 8, 2016, https://ffrf.org/publications/freethought-today/item/28552-evolution-and-atheism-best-friends-forever-jerry-coyne.

32. Coyne, "Evolution and Atheism."

33. Richard Dawkins, *The God Delusion* (London: Bantam, 2006), 28.

34. Ronald S. Hermann, "On the Legal Issues of Teaching Evolution in Public Schools," *American Biology Teacher* 75, no. 8 (October 2013): 539–43.

35. Richard Dawkins, "Stop the Anti-Vaccine Gospel," Richard Dawkins Foundation for Reason & Science, September 11, 2013, (Accessed February 2021. This page has since been removed from the site.)

36. Lynn Vincentnathan, S. Georg Vincentnathan, and Nicholas Smith, "Catholics and Climate Change Skepticism," *Worldviews* 20, no. 2 (2016): 125–49.

37. Coyne, "Evolution and Atheism."

38. Colin Schultz, "The Pope Would Like You to Accept Evolution and the Big Bang," *Smithsonian Magazine*, October 28, 2014, https://www.smithsonianmag.com/smart-news/pope-would-you-accept-evolution-and-big-bang-180953166/.

39. M. Elizabeth Barnes et al, "'Accepting Evolution Means You Can't Believe in God': Atheistic Perceptions of Evolution Among College Biology Students," *CBE Life Sciences Education* 19, no. 2 (June 1, 2020): 21.

40. Thomas H. McCall, "Will the Real Adam Please Stand Up? The Surprising Theology of Universal Ancestry," BioLogos, March 23, 2020, https://biologos.org/articles/series/book-review-the-genealogical-adam-and-eve/will-the-real-adam-please-stand-up-the-surprising-theology-of-universal-ancestry.

41. Andrew Dickson White, *A History of the Warfare of Science with Theology in Christendom*, 2 vols. (New York: D. Appleton and Co., 1896), 1.v.iii.

42. Keith S. Taber et al., "To What Extent Do Pupils Perceive Science to Be Inconsistent with Religious Faith? An Exploratory Survey of 13–14 year-old English Pupils," *Science Education International* 22, no. 2 (June 2011): 99–118.

43. John Theodore Houghton, *Does God Play Dice?* (Downers Grove, IL: Inter-Varsity Press, 1988), 40–41.

44. Houghton, *Does God Play Dice?*

45. Andrew Freedman, "John Houghton, Renowned Climate Scientist Who Led IPCC Reports, Dies of Coronavirus at 88," *Washington Post*, April 21, 2020, https://www.washingtonpost.com/local/obituaries/john-houghton-renowned-climate-scientist-who-led-ipcc-reports-dies-of-coronavirus-at-88/2020/04/20/c6b6819c-81ab-11ea-a3ee-13e1ae0a3571_story.html.

46. John Houghton, "Global Warming Is Now a Weapon of Mass Destruction," *The Guardian*, July 28, 2003.

47. The award's website is https://www.consejoculturalmundial.org/awards/world-award-of-science/.

48. Her website is http://www.katharinehayhoe.com/wp2016/biography/.

49. Darren Devine, "Teacher Inspired Nobel Prize-Winner Sir John Houghton," *Wales Online* March 29, 2013, https://www.walesonline.co.uk/news/local-news/teacher-inspired-nobel-prize-winner-sir-2210911

50. John Dickson, *Good Earth: Undeceptions*, Season 3 (podcast), https://undeceptions.com/podcast/good-earth.

51. Full transcript at Deborah Haarsma, "Seeing God in Everyday Work," BioLogos, July 7, 2014, https://biologos.org/articles/seeing-god-in-everyday-work.

52. https://biologos.org/about-us#our-mission.

53. https://community.dur.ac.uk/christianleadership.science/the-project/scientists-in-congregations/.

54. Personal correspondence.

55. Personal correspondence.

56. S. Joshua Swamidass, *The Genealogical Adam and Eve* (Downers Grove, IL: IVP Academic, 2019); the original article is at https://asa3.org/ASA/PSCF/2018/PSCF3-18Swamidass.pdf?fbclid=IwAR3JAgQrNWUV3NBmYdDQU7b2vhp0 UWQwnvRUss41jvPrDwJJMHSJw-9wY1I; further discussion is at https:// henrycenter.tiu.edu/2017/06/a-genealogical-adam-and-eve-in-evolution/?fbcli d=IwAR1ke4nzcb8EvqOLIcosAnPnim6PvaoKevjYV2hjMd-KyV5zNG2HN7fkb4c and https://henrycenter.tiu.edu/2020/08/the-genealogical-adam-and-eve-a-rejoinder/?fbc lid=IwAR3vpZY5R7uLNdKFx391AMafK3194emmO4 YffeEj0bSJQihZRrC9CBs52EU.

57. Nathan H. Lents, "Upcoming Book Leaves Scientific Possibility for Existence of 'Adam and Eve,' " *USA Today*, Opinion, October 4, 2019, https://eu.usatoday.com/ story/opinion/ 2019/10/04/upcoming-book-leaves- scientific-possibility-existence-adam-eve-column/ 3826195002/.

58. https://peacefulscience.org/mission-and-values/.

59. Personal correspondence.

60. Hardin, Numbers, and Binzley, *The Warfare Between Science and Religion*, 4.

61. Personal correspondence.

62. As quoted on the splashpage of the website: https://historyforatheists.com/.

63. Elaine Howard Ecklund et al., "Religion Among Scientists in International Context: A New Study of Scientists in Eight Regions," *Socius: Sociological Research for a Dynamic World* 2 (2016), https://doi.org/10.1177/2378023116664353.

64. Personal correspondence.

65. Rebecca McLaughlin, *Confronting Christianity* (Wheaton, IL: Crossway, 2019), 118.

# Selected Bibliography

## Primary Sources

The letters and other manuscript material cited in this book can be found in archives at the institutions listed below.

John William Draper Family Papers, Manuscript Division, Library of Congress, Washington, DC.

Andrew Dickson White Papers at Cornell University, Library Division of Rare and Manuscript Collections, 2B Carl A. Kroch Library, Cornell University, Ithaca, New York.

### Books and Pamphlets

*Account of the Proceedings at the Inauguration October 7th 1868.* Ithaca, NY: University Press, 1868.

*Addresses at the Thirtieth Meeting of the Unitarian Club of California, Held at San Francisco, Cal. April 26, 1897.* San Francisco: C. A. Murdock & Co, 1897.

*Annual Report of the American Institute of the City of New York, 1869–70.* Albany, NY: Argus Co., 1870.

Arnold, Matthew. *God & the Bible: A Review of Objections to "Literature & Dogma."* New York: Macmillan and Co., 1875.

Arnold, Matthew. *Literature & Dogma: An Essay Towards a Better Apprehension of the Bible.* New York: Macmillan and Co., 1874.

Arnold, Matthew. *St. Paul and Protestantism; with an Essay on Puritanism and the Church of England.* London: Smith, Elder & Co., 1870.

Barker, George F. *Memoir of John William Draper, 1811–1882: Read Before the National Academy, April 21, 1886.* Washington, DC: National Academy of Sciences, 1886.

*Catalogue of the Historical Library of Andrew Dickson White: The Protestant Reformation and Its Forerunners.* Ithaca, NY: Cornell University Press, 1889.

Chambers, Robert. *Explanations: A Sequel to Vestiges of the Natural History of Creation.* New York: Wiley and Putman, 1846.

Chambers, Robert. *Vestiges of the Natural History of Creation.* London: John Churchill, 1844.

Clark, John Spencer, ed. *The Life and Letters of John Fiske*, 2 vols. Boston: Houghton Mifflin Co., 1917.

Colenso, John William. *The Pentateuch and Book of Joshua Critically Examined.* New York: D. Appleton and Co., 1863.

Combe, George. *On the Relation Between Science and Religion.* Edinburgh: Maclachlan and Stewart, 1857.

Combe, George. *The Constitution of Man Considered in Relation to External Objects.* Boston: Carter and Hendee, 1829.

Cornell, Alonzo. *True and Firm: Biography of Ezra Cornell.* New York: A. S. Barnes & Co., 1884.

Cornell University. *First General Announcement*. Albany: Weed, Parsons and Co., 1868.

Derby, James Cephas. *Fifty Years Among Authors: Books and Publishers*. New York: C. W. Carleton & Co., 1884.

Draper, John William. *A History of the Intellectual Development of Europe*. New York: Harper & Brothers, 1863.

Draper, John William. *A Text-Book on Chemistry: For the Use of Schools and Colleges*. New York: Harper & Brothers, 1846.

Draper, John William. *A Text-Book on Natural Philosophy: For the Use of Schools and Colleges*. New York: Harper & Brothers, 1847.

Draper, John William. *History of the American Civil War*, 3 vols. New York: Harper & Brothers, 1867–1870.

Draper, John William. *History of the Conflict Between Religion and Science*. New York: D. Appleton & Co., 1874.

Draper, John William. *Human Physiology, Statical and Dynamical; or, The Conditions and Course of the Life of Man*. New York: Harper & Brothers, 1856.

Draper, John William. *Introductory Lecture in the Course of Chemistry*. New York: Hopkins & Jennings, 1841.

Draper, John William. *Introductory Lecture on Oxygen Gas*. New York: Joseph H. Jennings, 1848.

Draper, John William. *Introductory Lecture to the Course of Chemistry: Relations and Nature of Water*. New York: University of New York, 1845–1846.

Draper, John William. *Introductory Lecture to the Course of Chemistry: Relations of Atmospheric Air to Animals and Plants*. New York: University of New York, 1844–45.

Draper, John William. *Scientific Memoirs: Being Experimental Contributions to a Knowledge of Radiant Energy*. New York: Harper & Brothers, 1878.

Draper, John William. *The Influences of Physical Agents on Life: Being an Introductory Lecture to the Course on Chemistry and Physiology*. New York: John A. Gray, 1850.

Draper, John William. *Thoughts on the Future Civil Policy of America*. New York: Harper & Brothers, 1865.

Draper, John William. *Treatise on the Forces Which Produce the Organization of Plants*. New York: Harper & Brothers, 1844.

Draper, Thomas Waln-Morgan. *The Drapers in America, Being a History and Genealogy of Those of That Name and Connection*. New York: John Polhemus Printing Co., 1892.

Duncan, David, ed. *The Life and Letters of Herbert Spencer*. London: Methuen & Co., 1908.

Featherstone, J. S. *A Tribute of Grateful Remembrance to the Memory of the Rev. John Christopher Draper, Late Superintendent of the Wesleyan Methodist Society in the Sheerness Circuit*. Sheerness, 1829.

Fisk, Ethel F., ed. *The Letters of John Fiske*. New York: Macmillan Company, 1940.

Fiske, John. *A Century of Science and Other Essays*. Boston: Houghton, Mifflin & Co., 1899.

Fiske, John. *Edward Livingston Youmans: Interpreter of Science for the People: A Sketch of His Life with Selections from His Published Writings and Extracts form His Correspondence with Spencer, Huxley, Tyndall, and Others*. New York: D. Appleton & Co., 1894.

Fiske, John. *Outlines of Cosmic Philosophy, Based on the Doctrine of Evolution, with Criticisms on the Positive Philosophy*, 2 vols. London: Macmillan & Co., 1874.

Fiske, John. *The Destiny of Man Viewed in the Life of His Origin*. Boston: Houghton, Mifflin & Co., 1893.

Fiske, John. *The Idea of God as Affected by Modern Knowledge*. Boston: Houghton, Mifflin & Co., 1887.

Fiske, John. *Through Nature to God*. Boston: Houghton, Mifflin & Co., 1899.

*Freedom and Fellowship in Religion: A Collection of Essays and Addresses, Edited by a Committee of the Free Religious Association*. Boston: Roberts Brothers, 1875.

Frothingham, Octavius Brooks. *Recollections and Impressions, 1822–1890*. New York: G. P. Putnam's Sons, 1891.

Froude, J. A. *Short Studies on Great Subjects*, 4 vols. New York: Charles Scribner's Sons, 1888.

Froude, J. A. *The Nemesis of Faith*. London: John Chapman, 1849.

Herschel, John F. W. *A Preliminary Discourse on the Study of Natural Philosophy*. London: Longman, 1830.

Higgins, W. M. *The Mosaical and Mineral Geologies, Illustrated and Compared*. London: John Scoble, 1832.

Huxley, Leonard. *Life and Letters of Thomas Henry Huxley*, 2 vols. New York: D. Appleton & Co., 1901.

Huxley, Thomas Henry. *Collected Essays*, 9 vols. New York: D. Appleton & Co., 1916.

Huxley, Thomas Henry. *Essays upon Some Controverted Questions*. New York: D. Appleton & Co., 1892.

Huxley, Thomas Henry. *Evidence as to Man's Place in Nature*. New York: D. Appleton & Co., 1863.

Huxley, Thomas Henry. *Lay Sermons, Addresses, and Reviews*. New York: D. Appleton & Co., 1870.

Lyell, Charles. *Principles of Geology: Being an Inquiry How Far the Former Changes of the Earth's Surface Are Referable to Causes Now in Operation*, 4 vols., 5th ed. London: John Murray, 1837.

Ogden, Robert Morris. *Diaries of Andrew D. White*. Ithaca, NY: Cornell University Press, 1959.

Oort, H., I. Hooykaas, and A. Kuenen. *The Bible for Learners*, 3 vols. Translated by Philip Henry Wicksteed. Boston: Roberts Brothers, 1878.

*Proceedings at the First Annual Meeting of the Free Religious Association, Held in Boston, May 28th and 29th, 1868*. Boston: Adams & Co., 1868.

*Proceedings at the Ninth Annual Meeting of the Free Religious Association, Held in Boston, June 1 and 2, 1876*. Boston: Free Religious Association, 1876.

*Proceedings of the First American Congress of Liberal Religious Societies, Held at Chicago, May 22, 23, 24 & 25, 1895*. Chicago: Bloch & Newman, 1894.

*Report of the Committee on Organization, Presented to the Trustees of the Cornell University. October 21st, 1866*. Albany, NY: C. Van Benthuysen & Sons, 1867.

Ripley, Dorothy. *The Extraordinary Conversion, and Religious Experience of Dorothy Ripley, with Her First Voyage and Travels in America*. New York: G. and R. Waite, 1810.

Sexton, George. *Scientific Materialism Calmly Considered: Being a Reply to the Address Delivered Before the British Association, at Belfast, on August 19th, 1874, by Professor Tyndall*. London: J. Burns, 1874.

Slugg, J. T. *Woodhouse Grove School: Memorials and Reminiscences*. London: A. Ireland and Co., 1885.

Spencer, Herbert. *An Autobiography*, 2 vols. New York: D. Appleton & Co., 1904.

Spencer, Herbert. *First Principles of a New System of Philosophy*. New York: D. Appleton & Co., 1864.

Stephen, Leslie. *History of English Thought in the Eighteenth Century*. 1876.

Tyndall, John. *Fragments of Science: A Series of Detached Essays, Addresses, and Reviews*, 2 vols. New York: D. Appleton & Co., 1898.

Wace, Henry. *Christianity and Agnosticism: Reviews of Some Recent Attacks on the Christian Faith*. London: Society for Promoting Christian Knowledge, 1905.

Ward, James. *Naturalism and Agnosticism*. New York: Macmillan Company, 1899.

Whewell, William. *History of the Inductive Sciences, from the Earliest to the Present Times*, 3 vols. London: John W. Parker, 1847.

White, Andrew Dickson. *A History of the Warfare of Science and Theology in Christendom*, 2 vols. New York: D. Appleton & Co., 1896.

White, Andrew Dickson. *Autobiography*, 2 vols. New York: Century Co., 1905.

White, Andrew Dickson. *European Schools of History and Politics*. Baltimore: N. Murray, 1887.

White, Andrew Dickson. *Evolution and Revolution: An Address Delivered at the Annual Commencement of the University of Michigan, June 26, 1890*. Ann Arbor: University of Michigan, 1890.

White, Andrew Dickson. *On Studies in General History and the History of Civilization*. New York: G. P. Putnam's Sons, 1885.

White, Andrew Dickson. *Outlines of Lectures on History, Addressed to the Students of the Cornell University*. Ithaca, NY: Cornell University Press, 1883.

White, Andrew Dickson. *Some Practical Influences of German Thought upon the United States*. Ithaca, NY: Andrus & Church, 1884.

White, Andrew Dickson. *The Message of the Nineteenth Century to the Twentieth*. New Haven, CT: Tuttle, Morehouse & Taylor, 1883.

White, Andrew Dickson. *The Most Bitter Foe of Nations, and the Way to Its Permanent Overthrow*. New Haven, CT: T. J. Stafford, 1866.

White, Andrew Dickson. *The Warfare of Science*. New York: D. Appleton & Co., 1876.

Youmans, Edward L. *A Class-Book of Chemistry, in Which the Principles of the Science Are Familiarly Explained and Applied to the Arts, Agriculture, Physiology, Dietetics, Ventilation, and the Most Important Phenomena of Nature: Designed for the Use of Academies and Schools, and for Popular Reading*. New York: D. Appleton & Co., 1857.

Youmans, Edward L., ed. *The Correlation and Conversation of Forces: A Series of Expositions, by Prof. Grove, Prof. Helmholtz, Dr. Mayer, Dr. Faraday, Prof. Liebig, and Dr. Carpenter, with an Introduction and Brief Biographical Notices of the Chief Promoters of the New Views*. New York: D. Appleton & Co., 1865.

Youmans, Edward L., ed. *The Culture Demanded by Modern Life; a Series of Addresses and Arguments on the Claims of Scientific Education*. New York: D. Appleton & Co., 1867.

## Secondary Sources

Altholz, Josef L. *Anatomy of a Controversy: The Debate Over "Essays and Reviews," 1860–1864*. Aldershot, UK: Scholar Press, 1994.

Altholz, Josef L. "The Mind of Victorian Orthodoxy: Anglican Responses to 'Essays and Reviews', 1860–1864." *Church History* 54, no. 2 (1982): 186–97.

Altschuler, Glenn C. *Andrew D. White—Educator, Historian, Diplomat*. Ithaca, NY: Cornell University Press, 1979.

Altschuler, Glenn C. "From Religion to Ethics: Andrew D. White and the Dilemma of a Christian Rationalist." *Church History* 47, no. 3 (September 1978): 308–24.

Arx, Jeffrey Paul von. *Progress and Pessimism: Religion, Politics, and History in Late Nineteenth Century Britain.* Cambridge, MA: Harvard University Press, 1985.

Barth, Karl. *Protestant Theology in the Nineteenth Century: Its Background and History.* London: SCM Press, 1972.

Barton, Ruth. *The X Club: Power and Authority in Victorian Science.* Chicago: University of Chicago Press, 2018.

Becker, Carl L. *Cornell University: Founders and the Founding.* Ithaca, NY: Cornell University Press, 1943.

Bellot, H. Hale. *University College London 1826–1926.* London: University of London Press, 1929.

Berry, R. J., ed. *The Lion Handbook of Science & Christianity.* Oxford: Lion Handbooks, 2012.

Bishop, Morris. *A History of Cornell.* Ithaca, NY: Cornell University Press, 1962.

Bowler, Peter J. *Evolution: The History of an Idea.* Berkeley: University of California Press, 2009.

Bowler, Peter J. *Reconciling Science and Religion: The Debate in Early-Twentieth-Century Britain.* Chicago: University of Chicago Press, 2001.

Bowler, Peter J., and Iwan Rhys Morus. *Making Modern Science: A Historical Survey.* Chicago and London: University of Chicago Press, 2005.

Brasch, Frederick E. "List of Foundation Members of the History of Science Society." *Isis* 7, no. 3 (1925): 371–93.

Brock, W. H., N. D. McMillan, and R. C. Mollan, eds. *John Tyndall: Essays on a Natural Philosopher.* Dublin: Royal Dublin Society, 1981.

Brooke, John Hedley. "Presidential Address: Does the History of Science Have a Future?" *British Journal for the History of Science* 32, no. 1 (1999): 1–20.

Brooke, John Hedley. *Science and Religion: Some Historical Perspectives.* Cambridge: Cambridge University Press, 1991.

Brooke, John Hedley, and Geoffrey Cantor. *Reconstructing Nature: The Engagement of Science and Religion: Glasgow Clifford Lectures.* Edinburgh: T&T Clark, 1998.

Brooke, John Hedley, and Ronald L. Numbers, eds. *Science and Religion Around the World.* Oxford: Oxford University Press, 2011.

Brooke, John Hedley, Margaret J. Osler, and Jitse Van der Meer, eds. "Science in Theistic Contexts: Cognitive Dimensions." *Osiris* 16 (2001).

Buckley, Jerome Hamilton. *The Triumph of Time: A Study of the Victorian Concepts of Time, History, Progress, and Decadence.* Cambridge, MA: Belknap Press of Harvard University Press, 1966.

Buckley, Michael J. *At the Origins of Modern Atheism.* New Haven, CT: Yale University Press, 1987.

Budd, Susan. *Varieties of Unbelief: Atheists and Agnostics in English Society 1850–1960.* New York: Holmes and Meier, 1977.

Burtt, E. A. *Metaphysical Foundations of Modern Physical Science: A Historical and Critical Essay.* New York: Kegan Paul, Trench, Trübner & Co., 1925.

Butler, Lance St. John. *Victorian Doubt: Literary and Cultural Discourses.* New York: Harvester and Wheatsheaf, 1990.

Butterfield, Herbert. *The Origins of Modern Science: 1300–1800.* New York: Macmillan Company, 1958.

Butterfield, Herbert. *The Whig Interpretation of History*. New York: W. W. Norton & Co., 1965.

Byrne, Peter. *Natural Religion and the Nature of Religion: The Legacy of Deism*. New York: Routledge, 1989.

Cahan, David, ed. *From Natural Philosophy to the Sciences: Writing the History of Nineteenth-Century Science*. Chicago: University of Chicago Press, 2003.

Cannon, Walter F. "John Herschel and the Idea of Science." *Journal of the History of Ideas* 22, no. 2 (1961): 215–39.

Cannon, Walter F. "Scientists and Broad Churchmen: An Early Victorian Intellectual Network." *Journal of British Studies* 4, no. 1 (1964): 65–88.

Cannon, Walter F. "The Problem of Miracles in the 1830s." *Victorian Studies* 4, no. 1 (1960): 4–32.

Cantor, Geoffrey. *Religion and the Great Exhibition of 1851*. New York: Oxford University Press, 2011.

Cantor, Geoffrey. "Science, Providence, and Progress at the Great Exhibition." *Isis* 103, no. 3 (September 2012): 439–59.

Carter, Paul A. *The Spiritual Crisis of the Gilded Age*. DeKalb: Northern Illinois University Press, 1971.

Cashdollar, Charles D. *The Transformation of Theology, 1830–1890: Positivism and Protestant Thought in Britain and America*. Princeton, NJ: Princeton University Press, 2014.

Chadwick, Owen. "Gibbon and the Church Historians." *Daedalus* 105, no. 3 (1976): 111–23.

Chadwick, Owen. *The Secularization of the European Mind in the 19th Century*. Cambridge: Cambridge University Press, 1975.

Chadwick, Owen. *The Victorian Church*, 2 vols. London: Adam & Charles Black, 1966–1970.

Chapman, Allan. *Slaying the Dragons: Destroying Myths in the History of Science and Faith*. Oxford: Lion, 2013.

Chew, Samuel C. *Fruit Among the Leaves*. New York: Appleton-Century-Crofts, 1950.

Clagett, Marshall. *Greek Science in Antiquity*. London: Abelard-Schuman, 1957.

Clagett, Marshall. *The Science of Mechanics in the Middle Ages*. Madison: University of Wisconsin Press, 1959.

Cohen, I. Bernard. "George Sarton." *Isis* 48, no. 3 (1957): 286–300.

Cohen, I. Bernard. *Revolution in Science*. Cambridge, MA: Belknap Press of Harvard University Press, 1985.

Cohen, I. Bernard. "The Eighteenth-Century Origins of the Concept of Scientific Revolution." *Journal of the History of Ideas* 37, no. 2 (1976): 257–88.

Cohen, I. Bernard. "The Isis Crises and the Coming of Age of the History of Science Society." *Isis* 90, Suppl. (1999): S28–S42.

Colish, Marcia L. *Medieval Foundation of the Western Intellectual Traditions, 400–1400*. New Haven, CT: Yale University Press, 2002.

Conant, James B. "George Sarton and Harvard University." *Isis* 48, no. 3 (1957): 301–05.

Cooke, Bill. "Joseph McCabe: A Forgotten Early Populariser of Science and Defender of Evolution." *Science and Education* 19 (2010): 461–64.

Cooke, Bill. *The Gathering of Infidels: A Hundred Years of the Rationalist Press Association*. Amherst, NY: Prometheus Books, 2004.

Copleston, Frederick. *A History of Philosophy*, 9 vols. New York: DoubleDay, 1994.

Corsi, Pietro. *Science and Religion: Baden Powell and the Anglican Debate, 1800–1860.* Cambridge: Cambridge University Press, 1988.

Cosslett, Tess. *Science and Religion in the Nineteenth Century.* Cambridge: Cambridge University Press, 1984.

Cosslett, Tess. *The "Scientific Movement" and Victorian Literature.* New York: St. Martin's Press, 1982.

Coyne, Jerry A. *Faith Versus Fact: Why Science and Religion Are Incompatible.* London: Penguin Books, 2016.

Cragg, Gerald R. *The Church and the Age of Reason, 1648–1789.* London: Penguin Books, 1990.

Crane, Ronald S. "Anglican Apologetics and the Idea of Progress: 1699–1745." *Modern Philology* 31, nos. 3 and 4 (1934): 273–306, 349–82.

Dale, Richard, ed. *The Scientific Achievement of the Middle Ages.* Philadelphia: University of Pennsylvania Press, 1973.

Davis, Edward B. "Newton's Rejection of the 'Newtonian World View': The Role of Divine Will in Newton's Natural Philosophy." *Science and Christian Belief* 3, no. 1 (1991): 103–17.

Dawkins, Richard. *Science in the Soul: Selected Writings of a Passionate Rationalist.* New York: Random House, 2017.

Dawkins, Richard. *The God Delusion.* Boston: Mariner Books, 2006.

Dawson, Christopher. *Progress and Religion: An Historical Inquiry.* London: Sheed and Ward, 1929.

Desmond, Adrian. *Archetypes and Ancestors: Paleontology in Victorian London, 1850–1875.* Chicago: University of Chicago Press, 1984.

Desmond, Adrian. *Huxley: From Devils' Disciple to Evolution's High Priest.* Reading, MA: Addison-Wesley, 1997.

Desmond, Adrian. "Redefining the X Axis: 'Professionals,' 'Amateurs,' and the Making of Mid-Victorian Biology—A Progress Report." *Journal of the History of Biology* 34, (2001): 3–50.

Desmond, Adrian. *The Politics of Evolution: Morphology, Medicine, and Reform in Radical London.* Chicago and London: University of Chicago Press, 1989.

Desmond, Adrian, and James Moore. *Darwin.* London: Michael Joseph, 1991.

DeWitt, Anne. *Moral Authority, Men of Science, and the Victorian Novel.* Cambridge: Cambridge University Press, 2013.

Dillenberger, John. *Protestant Thought and Natural Science: A Historical Interpretation.* London: Collins, 1961.

Dixon, Thomas. *Science and Religion: A Very Short Introduction.* Oxford: Oxford University Press, 2008.

Dixon, Thomas, Geoffrey Cantor, and Stephen Pumfrey, eds. *Science and Religion: New Historical Perspectives.* Cambridge: Cambridge University Press, 2010.

Dorf, Philip. *The Builder: A Biography of Ezra Cornell.* New York: Macmillan Company, 1952.

Eisely, Loren C. *Darwin's Century: Evolution and the Men Who Discovered It.* London: Gollancz, 1958.

Eisen, Sydney. "Frederic Harrison and Herbert Spencer: Embattled Unbelievers." *Victorian Studies* 12, no. 1 (1968): 33–56.

Eisen, Sydney, and Bernard V. Lightman, eds. *Victorian Science and Religion: A Bibliography with Emphasis on Evolution, Belief, and Unbelief, Comprised of Works Published from c. 1900-1975*. Hamden, CT: Archon Books, 1984.

Ellegård, Alvar. *Darwin and the General Reader: The Reception of Darwin's Theory of Evolution in the British Periodical Press, 1859-1872*. Chicago and London: University of Chicago Press, 1990.

Ellegård, Alvar. *The Readership of the Periodical Press in Mid-Victorian Britain*. Göteborg, Sweden: Almqvist &Wiksell Stockholm, 1957.

Ellis, Ieuan. *Seven Against Christ: A Study of "Essays and Reviews."* Leiden, The Netherlands: Brill, 1980.

Ferngren, Gary B., ed. *Science and Religion: A Historical Introduction*. Baltimore and London: Johns Hopkins University Press, 2002.

Ferngren, Gary B., ed. *The History of Science and Religion in the Western Tradition: An Encyclopedia*. New York: Garland, 2000.

Fleming, Donald. *John William Draper and the Religion of Science*. Philadelphia: University of Pennsylvania Press, 1950.

"Focus: 100 Volumes of *Isis*: The Vision of George Sarton." *Isis* 100, no. 1 (2009): 58-107.

Foster, Michael B. "Christian Theology and Modern Science of Nature, Part 1." *Mind* 44, no. 176 (1935): 439-66.

Foster, Michael B. "Christian Theology and Modern Science of Nature, Part 2." *Mind* 45, no. 177 (1936): 1-27.

Foster, Michael B. "The Christian Doctrine of Creation and the Rise of Modern Natural Science." *Mind* 43, no. 172 (1934): 446-68.

Francis, Mark. *Herbert Spencer and the Invention of Modern Life*. New York: Cornell University Press, 2007.

Frängsmyr, Tore. "Science of History: George Sarton and the Positivist Tradition in the History of Science." *Lychnos* 74 (1973-1974): 104-44.

Funkenstein, Amos. *Theology and the Scientific Imagination: From the Middle Ages to the Seventeenth Century*. Princeton, NJ: Princeton University Press, 1986.

Gascoigne, John. *Cambridge in the Age of the Enlightenment: Science, Religion, and Politics from the Restoration to the French Revolution*. Cambridge: Cambridge University Press, 1989.

Gascoigne, John. "From Bentley to the Victorians: The Rise and Fall of British Newtonian Natural Theology." *Science in Context* 2, no. 2 (1988): 219-56.

Gay, Peter. *The Enlightenment*, 2 vols. New York and London: W. W. Norton, 1995.

"George Sarton Memorial Issue." *Isis* 48, no. (1957): 281-350.

Gillespie, Michael Allen. *The Theological Origins of Modernity*. Chicago: University of Chicago Press, 2008.

Gingerich, Owen. *God's Planet*. Cambridge, MA: Harvard University Press, 2014.

Good, H. G. "Edward L. Youmans: A National Teacher of Science." *Scientific Monthly* 18, no. 3 (March 1924): 306-17.

Grant, Edward. *A History of Natural Philosophy: From the Ancient World to the Nineteenth Century*. Cambridge: Cambridge University Press, 2007.

Grant, Edward. *Science & Religion, 400 BC-AD 1550: From Aristotle to Copernicus*. Baltimore: Johns Hopkins University Press, 2004.

Grant, Edward. *The Foundation of Modern Science in the Middle Ages: Their Religious, Institutional, and Intellectual Contexts*. Cambridge: Cambridge University Press, 1996.

Gregory, Brad S. *The Unintended Reformation: How a Religious Revolution Secularized Society*. Cambridge, MA: Belknap Press of Harvard University Press, 2012.

Haar, Charles M. "E.L. Youmans: A Chapter in the Diffusion of Science in America." *Journal of the History of Ideas* 9, no. 2 (1948): 193–213.

Hannam, James. *God's Philosophers: How the Medieval World Laid the Foundations of Modern Science*. London: Icon Books, 2009.

Hardin, Jeff, Ronald L. Numbers, and Ronald A. Binzley, eds. *The Warfare Between Science and Religion: The Idea That Wouldn't Die*. Baltimore: Johns Hopkins University Press, 2018.

Harris, Sam. *Letter to a Christian Nation*. New York: Alfred A. Knopf, 2006.

Harris, Sam. *The End of Faith: Religion, Terror, and the Future of Reason*. New York: W. W. Norton, 2004.

Harrison, Peter. *"Religion" and the Religions in the English Enlightenment*. New York: Cambridge University Press, 1990.

Harrison, Peter. " 'Science' and 'Religion': Constructing the Boundaries." *Journal of Religion* 86 (2006): 81–106.

Harrison, Peter. "Sentiments of Devotion and Experimental Philosophy in Seventeenth-Century England." *Journal of Medieval and Early Modern Studies* 44, no. 1 (2014): 113–33.

Harrison, Peter. *The Bible, Protestantism, and the Rise of Natural Science*. New York: Cambridge University Press, 1998.

Harrison, Peter, ed. *The Cambridge Companion to Science and Religion*. Cambridge: Cambridge University Press, 2010.

Harrison, Peter. *The Fall of Man and the Foundations of Science*. New York: Cambridge University Press, 2007.

Harrison, Peter. *The Territories of Science and Religion*. Chicago: University of Chicago Press, 2015.

Harrison, Peter, Ronald L. Numbers, and Michael H. Shank, eds. *Wrestling with Nature: From Omens to Science*. Chicago: University of Chicago Press, 2011.

Hart, David Bentley. *Atheist Delusions: The Christian Revolution and Its Fashionable Enemies*. New Haven, CT: Yale University Press, 2009.

Haught, John F. *God and the New Atheism: A Critical Response to Dawkins, Harris, and Hitchens*. Louisville, KY: Westminster John Knox Press, 2008.

Hearnshaw, F. J. C. *The Centenary History of King's College London, 1828–1928*. London: George G. Harrap & Co., 1926.

Henry, John. *A Short History of Scientific Thought*. New York: Palgrave Macmillan, 2012.

Henry, John. *The Scientific Revolution and the Origins of Modern Science*. New York: Palgrave Macmillan, 2008.

Hesketh, Ian. *Of Apes and Ancestors: Evolution, Christianity, and the Oxford Debate*. Toronto and London: University of Toronto Press, 2009.

Hindmarsh, D. Bruce. *The Spirit of Early Evangelicalism: True Religion in a Modern World*. Oxford: Oxford University Press, 2018.

Hilts, Victor L. "History of Science at the University of Wisconsin." *Isis* 75, no. 1 (1984): 63–94.

Hitchens, Christopher. *God Is not Great: How Religion Poisons Everything*. New York: Hachette Book Group, 2007.

Hooykaas, R. *Religion and the Rise of Modern Science*. Edinburgh: Scottish Academic Press, 1972.

Hutchison, William R. *The Modernist Impulse in American Protestantism*. Durham, NC: Duke University Press, 1992.

Hyman, Gavin. *A Short History of Atheism*. London and New York: I. B. Tauris, 2010.

Irvine, William. *Apes, Angels, and Victorians: The Story of Darwin, Huxley, and Evolution*. New York: McGraw-Hill Book Co., 1955.

Jacob, Margaret C. *The Newtonians and the English Revolution 1689–1720*. Hassocks, UK: Harvester Press, 1976.

Jenson, J. Vernon. "The X Club: Fraternity of Victorian Scientists." *British Journal for the History of Science* 5, no. 1 (1970): 63–72.

Jones, Tod E. *The Broad Church: A Biography of a Movement*. Lanham, MD: Lexington Books, 2003.

Jordan, Philip D. *The Evangelical Alliance for the United States of America, 1847–1900: Ecumenism, Identity and the Religion of the Republic*. New York: Edwin Mellon, 1982.

Kim, Stephen S. *John Tyndall's Transcendental Materialism and the Conflict Between Religion and Science in Victorian England*. Lewiston: Mellen University Press, 1996.

Knight, David. *Science and Spirituality: The Volatile Connection*. London and New York: Routledge, 2004.

Knight, David. *Sources for the History of Science 1660–1914*. Cambridge: Cambridge University Press, 1976.

Knight, David M., and Matthew D. Eddy, eds. *Science and Beliefs: From Natural Philosophy to Natural Science, 1700–1900*. Aldershot, UK: Ashgate, 2005.

Kocher, Paul H. *Science and Religion in Elizabethan England*. San Marino, CA: Huntington Library, 1953.

Lane, Christopher. *The Age of Doubt: Tracing the Roots of our Religious Uncertainty*. New Haven, CT: Yale University Press, 2011.

Larsen, Timothy. *A People of One Book: The Bible and the Victorians*. New York: Oxford University Press, 2011.

Larsen, Timothy. *Contested Christianity: The Political and Social Contexts of Victorian Theology*. Waco, TX: Baylor University Press, 2004.

Larsen, Timothy. *Crisis of Doubt: Honest Faith in Nineteenth-Century England*. New York: Oxford University Press, 2008.

Larson, Edward. *Summer for the Gods: The Scopes Trial and America's Continuing Debate Over Science and Religion*. New York: Basic Books, 1997.

Lash, Nicholas. *The Beginning and the End of "Religion."* New York: Cambridge University Press, 1996.

Leverette, William, Jr. "E.L. Youmans' Crusade for Scientific Autonomy and Respectability." *American Quarterly* 1 (1965): 12–32.

Leverette, William, Jr. "Science and Values: A Study of Edward L. Youmans' *Popular Science Monthly*, 1872–1887." PhD diss., Vanderbilt University, 1963.

Lightman, Bernard. "Christian Evolutionists in the United States, 1860–1900." *Journal of Cambridge Studies* 4, no. 4 (2009): 14–22.

Lightman, Bernard. "Does the History of Science and Religion Change Depending on the Narrator? Some Atheist and Agnostic Perspectives." *Science and Christian Belief* 24 (2012): 149–68.

Lightman, Bernard. *Evolutionary Naturalism in Victorian Britain: The "Darwinians" and Their Critics*. Aldershot, UK: Ashgate, 2009.

Lightman, Bernard, ed. *Global Spencerism: The Communication and Appropriation of a British Evolutionist.* Leiden, The Netherlands: Brill, 2016.

Lightman, Bernard. "The International Scientific Series and the Communication of Darwinism." *Journal of Cambridge Studies* 5, no. 4 (2010): 27–38.

Lightman, Bernard. *The Origins of Agnosticism: Victorian Unbelief and the Limits of Knowledge.* Baltimore: Johns Hopkins University Press, 1987.

Lightman, Bernard. *Victorian Popularizers of Science: Designing Nature for New Audiences.* Chicago and London. University of Chicago Press, 2007.

Lightman, Bernard, ed. *Victorian Science in Context.* Chicago: University of Chicago Press, 1997.

Lightman, Bernard. "Victorian Sciences and Religions: Discordant Harmonies." *Osiris* 16 (2001): 343–66.

Lightman, Bernard, and Michael S. Reidy, eds. *The Age of Scientific Naturalism: Tyndall and His Contemporaries.* London: Pickering & Chatto, 2014.

Lindberg, David C. *The Beginnings of Western Science: The European Scientific Tradition in Philosophical, Religious, and Institutional Context, 600 B.C. to A.D. 1450.* Chicago: University of Chicago Press, 1992.

Lindberg, David C., and Ronald L. Numbers. "Beyond War and Peace: A Reappraisal of the Encounter Between Christianity and Science." *Church History* 55 (1986): 338–54.

Lindberg, David C., and Ronald L. Numbers, eds. *God and Nature: Historical Essays on the Encounter Between Christianity and Science.* Berkeley: University of California Press, 1986.

Lindberg, David C., and Ronald L. Numbers, eds. *When Science & Christianity Meet.* Chicago and London: University of Chicago Press, 2003.

Livingstone, David N. *Adam's Ancestors: Race, Religion, and the Politics of Human Origins.* Baltimore: Johns Hopkins University Press, 2008.

Livingstone, David N. *Darwin's Forgotten Defenders: The Encounter Between Evangelical Theology and Evolutionary Thought.* Grand Rapids, MI: Eerdmans, 1987.

Livingstone, David N. *Putting Science in Its Place: Geographies of Scientific Knowledge.* Chicago: University of Chicago Press, 2003.

Livingstone, David N. "Science and Religion: Towards a New Cartography." *Christian Scholar's Review* 26, no. 3 (Spring 1997): 270–92.

Livingstone, David N., D. G. Hart, and Mark A. Noll, eds. *Evangelicals and Science in Historical Perspective.* New York: Oxford University Press, 1999.

MacLeod, Roy M. *The "Creed of Science" in Victorian England.* Aldershot, UK: Ashgate, 2000.

MacLeod, Roy M. "The X-Club: A Social Network of Science in Late-Victorian England." *Notes and Records of the Royal Society of London* 24, no. 2 (1970): 305–22.

Madison, Charles A. *Book Publishing in America.* New York: McGraw-Hill Book Company, 1966.

Marjorie, Wheeler-Barclay. *The Science of Religion in Britain, 1860–1915.* Charlottesville: University of Virginia Press, 2010.

Marsden, George M. *Fundamentalism and American Culture.* New York: Oxford University Press, 1980.

Marsden, George M. *The Soul of the American University: From Protestant Establishment to Established Nonbelief.* New York: Oxford University Press, 1994.

Meadows, A. J., ed. *Development of Science Publishing in Europe.* Amsterdam: Elsevier Science Publishers, 1980.

Merton, Robert K. "George Sarton: Episodic Recollection by an Unruly Apprentice." *Isis* 76, no. 4 (1985): 470–86.

Merton, Robert K. "Puritanism, Pietism, and Science." *The Sociological Review* 28 (1936): 1–30.

Merton, Robert K. *Science, Technology and Society in Seventeenth-Century England.* New York: Harper & Row, 1970.

Merton, Robert K. "Science, Technology and Society in Seventeenth Century England." *Osiris* 4 (1938): 360–632.

McLeod, Hugh. *European Religion in the Age of the Great Cities, 1830–1930.* London: Routledge, 1995.

McLeod, Hugh. *Secularisation in Western Europe, 1848–1914.* New York: St. Martin's Press, 2000.

Moore, James R. "Evangelicals and Evolution: Henry Drummond, Herbert Spencer, and the Naturalisation of the Spiritual World." *Scottish Journal of Theology* 38, no. 3 (1985): 383–418.

Moore, James R., ed. *History, Humanity and Evolution: Essays for John C. Greene.* Cambridge: Cambridge University Press, 1989.

Moore, James R. *The Post-Darwinian Controversies: A Study of the Protestant Struggle to Come to Terms with Darwin in Great Britain and America 1879–1900.* Cambridge: Cambridge University Press, 1979.

Morgan, Charles. *The House of Macmillan (1843–1943).* London: Macmillan, 1943.

Numbers, Ronald L. *Darwinism Comes to America.* Cambridge, MA: Harvard University Press, 1998.

Numbers, Ronald L., ed. *Galileo Goes to Jail and Other Myths About Science and Religion.* Cambridge, MA: Harvard University Press, 2009.

Numbers, Ronald L. "Science and Religion." *Osiris* (2nd Series) 1 (1985): 59–80.

Numbers, Ronald L. "The American History of the Science Society or the International History of Science Society? The Fate of Cosmopolitanism Since George Sarton." *Isis* 100, no. 1 (2009): 103–07.

Numbers, Ronald L., and Kostas Kampourakis, eds. *Newton's Apple and Other Myths About Science.* Cambridge, MA: Harvard University Press, 2015.

Olson, Richard G. *Science & Religion, 1450–1900: From Copernicus to Darwin.* Baltimore: Johns Hopkins University Press, 2004.

Olson, Richard G. *Science and Scientism in Nineteenth-Century Europe.* Urbana and Chicago: University of Illinois Press, 2007.

Osler, Margaret J., ed. *Reconfiguring the World: Nature, God, and Human Understanding from the Middle Ages to Early Modern Europe.* Baltimore: Johns Hopkins University Press, 2010.

Osler, Margaret J., ed. *Rethinking the Scientific Revolution.* Cambridge: Cambridge University Press, 2000.

Osler, Margaret J., and Paul Lawrence Farber, eds. *Religion, Science, and Worldview: Essays in Honor of Richard S. Westfall.* New York: Cambridge University Press, 1985.

Overton, Grant. *Portrait of a Publisher, and the First Hundred Years of the House of Appleton 1825–1925.* New York: D. Appleton & Co., 1925.

Painter, Borden. *The New Atheist Denial of History.* New York: Palgrave Macmillan, 2016.

Paradis, James G. *T. H. Huxley: Man's Place in Nature.* Lincoln: University of Nebraska Press, 1978.

Pearcey, Nancy R., and Charles B. Thaxton. *The Soul of Science: Christian Faith and Natural Philosophy*. Wheaton, IL: Crossway Books, 1994.

Plantinga, Alvin. *Where the Conflict Really Lies: Science, Religion, and Naturalism*. Oxford: Oxford University Press, 2011.

Raven, Charles E. *Science, Religion, and the Future: A Course of Eight Lectures*. New York: Macmillan Company, 1944.

Reardon, Bernard M. G. *Liberal Protestantism*. Stanford, CA: Stanford University Press, 1968.

Reardon, Bernard M. G. *Religion in the Age of Romanticism: Studies in Early Nineteenth-Century Thought*. New York: Cambridge University Press, 1985.

Reardon, Bernard M. G. *Religious Thought in the Nineteenth Century: Illustrated from Writers of the Period*. Cambridge: Cambridge University Press, 1966.

Reuben, Julie A. *The Making of the Modern University: Intellectual Transformation and the Marginalization of Morality*. Chicago: University of Chicago Press, 1996.

Richardson, W. Mark, and Wesley J. Wildman, eds. *Religion and Science: History, Method, Dialogue*. New York: Routledge, 1996.

Roberts, Jon H. *Darwinism and the Divine in America: Protestant Intellectuals and Organic Evolution, 1859–1900*. Madison: University of Wisconsin Press, 1988.

Roberts, Jon H. and James Turner. *The Sacred and the Secular University*. Princeton, NJ: Princeton University Press, 2000.

Rogers, Walter P. *Andrew D. White and the Modern University*. Ithaca, NY: Cornell University Press, 1942.

Ruse, Michael, and Robert J. Richards, eds. *The Cambridge Companion to the "Origin of Species."* Cambridge: Cambridge University Press, 2009.

Russell, Colin A. *Cross-Currents: Interactions Between Science and Faith*. Grand Rapids, MI: Eerdmans, 1985.

Russell, Colin A. "The Conflict Metaphor and Its Social Origins." *Science and Christian Belief* 1, no. 1 (1989): 3–26.

Russell, Colin A., R. Hooykaas, and David C. Goodman. *The "Conflict Thesis" and Cosmology*. Milton Keynes, UK: Open University Press, 1974.

Russell, Jeffrey Burton. *Inventing the Flat Earth: Columbus and Modern Historians*. Westport, CT, and London: Praeger, 1997.

Ryan, Robert M. *The Romantic Reformation: Religious Politics in English Literature, 1789–1824*. Cambridge: Cambridge University Press, 1997.

"Sarton, Science, and History." *Isis* 75, no. 1 (1984): 1–104.

Sarton, George. *Horus: A Guide to the History of Science: A First Guide for the Study of the History of Science with Introductory Essays on Science and Tradition*. Waltham, MA: Chronica Botanica Company, 1952.

Sarton, George. "Auguste Comte, Historian of Science: With a Short Digression on Clotilde de Vaux and Harriet Taylor." *Osiris* 10 (1952): 328–57.

Sarton, George. *Introduction to the History of Science*, Vol. 1: *From Homer to Omar Kahayyam*. Baltimore: Williams & Wilkins Company, 1927.

Sarton, George. "L'histoire de la science." *Isis* 1, no. 1 (1913): 3–46.

Sarton, George. "The Faith of the Humanist." *Isis* 3, no. 1 (1920): 3–6.

Sarton, George. "The History of Science." *Monist* 26, no. 3 (1916): 321–65.

Sarton, George. *The History of Science and the New Humanism*. New York: George Braziller, 1956.

Sarton, George. "The New Humanism." *Isis* 6, no. 1 (1924): 9–42.

Sarton, George. "The Teaching of the History of Science." *Isis* 4, no. (1921): 225–49.

Sarton, George. "The Teaching of the History of Science." *Isis* 13, no. 2 (1930): 272–97.

Sarton, George. "The Teaching of the History of Science." *Scientific Monthly* 7, no. 13 (1918): 193–211.

Sarton, Mary. *I Knew a Phoenix: Sketches for an Autobiography*. New York: W. W. Norton & Company, 1995.

Schaefer, Richard. "Andrew Dickson White and the History of Religious Future." *Zygon* 50, no. 1 (2015): 7–27.

Schlossberg, Herbert. *Conflict and Crisis in the Religious Life of Late Victorian England*. New Brunswick, NJ: Transaction Publishers, 2009.

Secord, James A. *Victorian Sensation: The Extraordinary Publication, Reception, and Secret Authorship of* Vestiges of the Natural History of Creation. Chicago: University of Chicago Press, 2000.

Secord, James A. *Visions of Science: Books and Readers at the Dawn of the Victorian Age*. Oxford: Oxford University Press, 2014.

Shapiro, B. J. "Latitudinarianism and Science in Seventeenth-Century England." *Past & Present* 40 (1968): 16–41.

Shea, Victor, and William Whitla, eds. *Essays and Reviews: The 1860 Text and Its Reading*. Charlottesville: University Press of Virginia, 2000.

Spellman, W. M. *The Latitudinarians and the Church of England, 1660–1700*. Athens: University of Georgia Press, 1993.

Stenger, Victor J. *God and the Folly of Faith: The Incompatibility of Science and Religion*. New York: Prometheus Books, 2012.

Stenger, Victor J. *God: The Failed Hypothesis—How Science Shows That God Does Not Exist*. New York: Prometheus Books, 2007.

Stenger, Victor J. *The New Atheism: Taking a Stand for Science and Reason*. New York: Prometheus Books, 2009.

Stump, J. B., ed. *Science and Christianity: An Introduction to the Issues*. Chichester, UK: Wiley-Blackwell, 2017.

Taylor, Charles. *A Secular Age*. Cambridge, MA: Belknap Press of Harvard University Press, 2007.

Taylor, Michael. *The Philosophy of Herbert Spencer*. London: Continuum, 2007.

Thackray, Arnold, and Robert K. Merton. "On Discipline Building: The Paradoxes of George Sarton." *Isis* 63, no. 4 (1972): 473–95.

Thomas, Keith. *Religion and the Decline of Magic: Studies in Popular Beliefs in Sixteenth Century England*. London: Weidenfeld & Nicolson, 1971.

Thorndike, Lynn. *History of Magic and Experimental Science*, 8 vols. New York: Columbia University Press, 1923–58.

Troeltsch, Ernst. *Protestantism and Progress: The Significance of Protestantism for the Rise of the Modern World*. Philadelphia: Fortress Press, 1986.

Turner, Frank Miller. *Between Science and Religion: The Reaction to Scientific Naturalism in Late Victorian England*. New Haven, CT: Yale University Press, 1974.

Turner, Frank Miller. *Contesting Cultural Authority: Essays in Victorian Intellectual Life*. Cambridge: Cambridge University Press, 1993.

Turner, Frank Miller. "The Victorian Conflict Between Science and Religion: A Professional Dimension." *Isis* 69, no. 3 (1978): 356–76.

Turner, Frank Miller. "Victorian Scientific Naturalism and Thomas Carlyle." *Victorian Studies* 18, no. 3 (1975): 325–43.

Turner, James. *Without God, Without Creed: The Origins of Unbelief in America*. Baltimore: Johns Hopkins University Press, 1985.

Ungureanu, James C. "A Yankee at Oxford: John William Draper at the British Association for the Advancement of Science at Oxford, 30 June 1860." *Notes and Records* 70, no. 2 (2016): 135–50.

Ungureanu, James C. "Edward L. Youmans and the 'Peacemakers' in Popular Science Monthly." *Fides et Historia* 51, no. 2 (Fall 2019): 13–32.

Ungureanu, James C. "Relocating the Conflict Between Science and Religion at the Foundations of the History of Science." *Zygon* 53, no. 4 (December 2018): 1106–30.

Ungureanu, James C. *Science, Religion, and the Protestant Tradition: Retracing the Origins of Conflict*. Pittsburgh: University of Pittsburgh Press, 2019.

Webb, George E. *The Evolution Controversy in America*. Lexington: University Press of Kentucky, 1994.

Weber, Max. *The Protestant Ethic and the Spirit of Capitalism*. Translated by Talcott Parsons. London: Routledge, 1930.

Webster, Charles. *The Great Instauration: Science, Medicine and Reform, 1626–1660*. London: Duckworth, 1975.

Welch, Claude. *Protestant Thought in the Nineteenth Century*, 2 vols. New Haven, CT: Yale University Press, 1972.

Westfall, Richard S. *Science and Religion in Seventeenth-Century England*. Hamden, CT: Archon Books, 1970.

White, Edward A. *Science and Religion in American Thought: The Impact of Naturalism*. Stanford, CA: Stanford University Press, 1952.

White, Paul. *Thomas Huxley: Making the "Man of Science."* Cambridge: Cambridge University Press, 2003.

Whitehead, Alfred North. *Science and the Modern World*. New York: Macmillan, 1925.

Wilson, S. Gordon. *The University of London and Its Colleges*. London: University Tutorial Press, 1923.

Wolfe, Gerard R. *The House of Appleton: The History of a Publishing House and Its Relationship to the Cultural, Social, and Political Events That Helped Shape the Destiny of New York City*. Metuchen, NJ: Scarecrow Press, 1981.

Wood, Paul., ed. *Science and Dissent in England, 1688–1945*. Aldershot, UK: Ashgate, 2004.

Woodhead, Linda, ed. *Reinventing Christianity: Nineteenth-Century Contexts*. Aldershot, UK: Ashgate, 2001.

Yeo, Richard. *Defining Science: William Whewell, Natural Knowledge, and Public Debate in Early Victorian Britain*. Cambridge: Cambridge University Press, 1993.

Yerxa, Donald A., ed. *Religion and Innovation: Antagonists or Partners?* London: Bloomsbury, 2016.

# Index